D1145302

THE HEROIC AGE
of
SCIENCE

The Conception, Ideals, and Methods of Science among the Ancient Greeks

by

WILLIAM ARTHUR HEIDEL

Research Professor of the Greek Language and Literature in Wesleyan University, Research Associate of the American Council of Learned Societies and of the Carnegie Institution of Washington

PUBLISHED FOR
CARNEGIE INSTITUTION OF WASHINGTON
BY
THE WILLIAMS & WILKINS COMPANY
Baltimore, 1933

CARNEGIE INSTITUTION OF WASHINGTON

PUBLICATION NO. 442

TABLE OF CONTENTS

The sages of early Greece form the heroic age of science. Like the first navigators in their own mythology, they boldly ventured their untried bark in a distant and arduous voyage, urged on by hopes of a supernatural success; and though they missed the imaginary golden prize which they sought, they unlocked the gates of distant regions, and opened the seas to the keels of the thousands of adventurers, who, in succeeding times, sailed to and fro, to the infinite increase of the mental treasures of mankind.

—WHEWELL, *History of the Inductive Sciences*, I, p. 68.

PREFACE

The subtitle of this modest booklet is intended to indicate its scope; but it can not adequately suggest the point of view from which it was written. Though that point of view is likely to call forth dissent and criticism, it was deliberately taken because it seemed the only right one in dealing with beginnings. Science now means many different things to different persons according to the special interest of each group; but, wherever one may turn, the name will be found to imply a quite definite subject and method. This is, of course, the result of specialization, which comes about partly through the limitations of personal interest and partly through the extension of the field of knowledge and the consequent realization of the practical impossibility of commanding a survey of the whole. In the beginning, however, there were neither established categories nor special methods and techniques; there was only the native intelligence schooled in practical affairs and directed to the various subjects which aroused the curiosity of the thinker. Men were surveying the scene, roughly sketching what they should like to achieve and forging the indispensable tools for its realization. Later generations have models ready to hand, on which they may improve according to their ability; the pioneer has at most the raw materials. But if he is limited in regard to the degree of refinement he may attain, he has by way of compensation a measure of freedom which later-comers hardly enjoy; and in building according to his circumstances he is perhaps fixing a type for all time. Such was the fortune of the earliest Greek thinkers; having in most fields of inquiry no models, they had perforce to depend upon their native resources and erect such structures as they could. If these fell short of the refinement that we now require, their work has at least the charm of showing the marks of their rough tools and revealing the means they employed in the building. The temple of science thus ceases to be a miracle, and becomes to one possessed of imagination an inspiration. But the science of the Greeks is not without its miracles; for in certain departments, such as mathematics and astronomy, the structures erected in the short time allotted to that people evoked the same degree of admiration for their simple beauty as do their masterpieces in literature, sculpture and architecture.

The purpose of the following discussion is to indicate how the Greeks set about the task of laying the foundations of science, rather than to recount their achievements. In order to do the latter, one must write a history of Greek science, a thing greatly to be desired, if it were possible. Strictly speaking, it is an impossibility; for, no matter what fortunate discoveries may be in store for us or for future generations, it is certain that for the crucial period from 600 to 100 B. C. the works of those who built the foundations are, with few exceptions, irretrievably lost. Some fields of inquiry, which were much cultivated, are represented by not a single original work, while others have left, as if to tantalize us, the merest fragments quoted by grammarians who had no interest in science. If one recalls that certain thinkers of the first rank either wrote not at all or exerted their influence chiefly by the spoken word, one recognizes the hopelessness of the task of the historian. At best he could offer a sketch pieced out with inferences more or less probable.

A quarter of a century ago the late Dr. R. S. Woodward, then President of Carnegie Institution of Washington, sought to enlist my assistance in the preparation of a co-operative work, which should take the place of Whewell's

History of the Inductive Sciences. Devoted as I had been for years to the study of the subject, the project naturally made a strong appeal to my interest; but I could not at the time be led to approve of it. Among other objections I urged my conviction that in the periods which laid the foundations of science—that of the ancient Greeks and that of the Renaissance—too little preliminary work had been done. Even now the situation is not materially changed. Whether I was responsible for the abandonment of the project, I do not know; if I were, I now regret it, because I have come to realize that it is given to no man, and to no age, to achieve perfection: we can at best carry on and report progress.

I hope later to make further contributions to the history of Greek science in various fields. The discussion which is herewith published is an attempt to exhibit by examples, chosen from a collection many years in the making, the way in which the Greeks thought of science and tried to solve its problems and to point out the relation between their procedure and the processes of the mind in practical affairs. The psychology involved has purposely been stated as simply as possible; some may regard it as superfluous.

The manuscript was completed three years ago. For the good fortune that it is published in these difficult times, I am indebted to Dr. Waldo G. Leland, Permanent Secretary of the American Council of Learned Societies, who has interested himself in the matter, and to Carnegie Institution of Washington, which has generously assumed the risks and included it in its list. To these and to the many friends, who have helped me in various ways, I offer my sincere thanks. I could wish that the book were more worthy of their kindness.

W. A. H.

Middletown, Conn., April 12, 1933

PART I. CONCEPTION AND IDEALS OF SCIENCE AMONG THE GREEKS

When one undertakes to survey ancient science, one meets at the very threshold a number of prejudices, which are calculated to deter the inquirer from prosecuting the study. One of these prejudices takes the form of a virtual denial that the ancient world had anything to show which might properly be called science.   Another arises from the confusion of science with philosophy.

That the Greeks developed science can be denied only if one defines the name in terms which are applicable to nothing but the most recent formulas.   It would be futile to attempt an answer to such a view; and the following discussion is intended to show how the Greeks understood science and how they faced the problems which the world presents to human intelligence.   As for the confusion of science and philosophy, there is indeed a difficulty, which is in part, however, of our own making.   Every historical account of ancient science begins with the views of thinkers who have from ancient times been called philosophers or, more properly, natural philosophers.   We are prone to forget that even a generation ago it was customary to give the name of Natural Philosophy to text-books which would now be entitled treatises on physics.   It happens to be my own opinion, though this is not the place to defend it, that the earliest Greek "philosophers" did not in any way trench upon the private domain of philosophy; but it is quite true that in certain outstanding treatises, such as the *Physics* of Aristotle, which for reasons historically intelligible have engaged the special attention of historians, there is far too much philosophy and far too little science in the modern

1

sense. If we were attempting a history of science, or of thought in general, it would be clearly our duty to explain the reasons for the prominence of such works in the conception which one frames of ancient science; for our present purpose it suffices to say that the reasons have little to do with the intrinsic merits of the particular works in question, but have very much to do with the temper of the late schools of ancient philosophy and the prevailing interests of the mediaeval schoolmen. The same influences which led to the exaltation of the purely speculative works of Aristotle operated to obscure and destroy many writings in the field of science, which, had they been preserved intact, would have fully redressed the balance and given the modern student, with his altered attitude, a fairer view of the achievements of the ancient world.

An unknown author,[1] probably of the First Century, sagely remarks that in literature critical judgment is the ultimate flowering of much experience; and this is obviously true of every sphere of life and thought. To the uncritical, things familiar are just what they are supposed to be, and are commonly regarded as having always been so; while that which is novel is assumed to be a new creation, having no more natural antecedents than the observer has preparatory ideas fitted to take in its essential character and significance. An historical sense is not congenital, but is developed only by long and intensive study; and yet it is indispensable to a larger view of any subject which has to do with man's intellectual life.

The Greeks were not blind to this fact. Though their own history was, and was known to them to be, short in comparison with that of other peoples, such as the Egyptians, their active intelligence reconstructed a long past, even picturing a succession of cosmic ages, punctuated by uni-

[1] Pseudo-Longinus, *De Sublimitate*, 6.

versal flood or conflagration, during which the arts of civilization were created and re-created. They ascribed the gift
of the arts to certain gods or to culture-heroes, such as
Palamedes or Prometheus, living long ago. They were well
aware that the arts which minister to the absolute necessities were discovered in pre-historic times; and Aristotle, who
had thought much about the evolution of human institutions and the interests from which they spring, several times
expressed the belief that all, or nearly all, the necessary arts
were achieved before men, now favored with leisure, turned
their attention to those which serve their hours of ease and
pastime.[2] Since he knew that the latter existed in the
earliest period of Greek history, he of course referred the former to pre-historic times. Among the occupations of leisure he reckoned science and philosophy.

In saying that the necessary arts had been achieved so
long ago, Aristotle doubtless meant no more than that they
had attained a stage of development sufficient to maintain
life under conditions then obtaining; for he could not be
unaware of many evidences of progress in historical times.
Some interesting observations are made by the author of
the treatise *On the Old School of Medicine:* "The art of medicine would never have been discovered to begin with, nor
would it have been sought—for there would have been no
need for medicine—if sick men had profited by the same
mode of living and regimen as the food, drink, and mode of
living of men in health, and if there had been no other things
for the sick better than these. But the fact is that sheer
necessity has caused men to seek and to discover medicine,
because sick men did not, and do not, profit by the same
regimen as do men in health. To trace the matter yet further back, I hold that not even the mode of living and
nourishment enjoyed at present by men in health would

[2] *Metaph.*, I. 1, 981 b 20 ff.; I. 2, 982 b 22 ff.

have been discovered, had a man been satisfied with the same food and drink as satisfy an ox, a horse, and every animal save man, for example the products of the earth— fruits, wood, and grass.   For on these they are nourished, grow, and live without pain, having no need at all of any other kind of living.   Yet I am of the opinion that in the beginning man also used this sort of nourishment.   Our present ways of living have, I think, been discovered and elaborated during a long period of time.   For many and terrible were the sufferings of men from strong and brutish diet, when they partook of food crude, uncompounded, and of strong qualities; just as men would suffer at the present day, falling into violent pains and diseases quickly followed by death.   Formerly indeed they probably suffered less, because they were used to it, but they suffered severely even then.   The majority naturally perished, having too weak a constitution, while the stronger resisted longer, just as at the present time some men easily deal with strong foods, while others do so only with many severe pains.   For this reason the ancients too seem to me to have sought for nourishment that harmonized with their constitution, and to have discovered that which we use now. . . . . Nevertheless the discovery was a great one, implying much investigation and art.   At any rate even at the present day those who study gymnastic and athletic exercises are constantly making some fresh discovery by investigating on the same method what food and what drink are best assimilated and make a man grow stronger."[3]

It is, no doubt, easy to realize that the essential means of satisfying the simpler material wants have come down to us from much cruder times; but it requires serious reflection to make us aware that much the same may be truthfully said of the furniture of our minds.   Our intellectual wants, like

[3] Hippocrates, *De Prisca Medicina*, 3–4, tr. Jones, adapted.

our physical requirements, are at bottom few, and the instruments needed to procure a first or rough satisfaction of them were forged long ago. Usener[4] took the view that a people's growth, and therefore its stock of fundamental ideas, is generally completed before it emerges into the light of history. There is much to be said for this view, though it must not be interpreted as denying the possibility or reality of developments and refinements in the ideas themselves or the shift of emphasis and the consequent rearrangement which they undergo. It is a matter of common observation that a given philosophy reflects a phase of social development which in reality has been outgrown. This may be explained on the assumption that it is only after ideas, loosed from their original associations and beginning to disintegrate, become fluid and move freely among themselves, that one begins to reflect on them, and to compose new patterns into which they enter as so many independent units. A wholly new idea is seldom, if ever, found; but even in the most advanced civilization there will always be discovered survivals of the most ancient notions which formed the original stock, because in this world of flux there is nothing so nearly permanent as certain ingrained ideas.

"One may say that primitive man has only religious apperceptive masses." "No matter what historical phenomenon we may trace to a remote past, we come at last to religion. All human conceptions, so far as they fall within the intellectual horizon of a pre-scientific age, have developed out of mythical conceptions; but religious ideas form the content, or at least the garb of myth."[5] One's judgment respecting the truth of these words will naturally depend on one's definition of myth and religion. What undoubtedly is true of primitive man is that, while he can not strictly be

[4] *Vorträge und Aufsätze*, 110 f.
[5] *Ibid.*, 43, 45.

said to regard the world as composed or controlled by persons, because "person" conveys a more clearly defined idea than he himself has framed, he looks upon all that occurs about and within him as the result of agencies akin to himself. The means he employs in the effort to control nature are essentially the same as those which he finds effective in dealing with his fellows. According to the stage which he attains in the organization of his world, his procedure may be described as magic or religion; but it is practically impossible, even at the highest stage yet reached by man, to divest the concept of causation, which underlies every effort at explaining the world, of its primal associations with agency. And when one speaks of the processes of nature the analogy with those which from the first held the interest of man is no less apparent.

This circumstance does not, of course, preclude insight, even very remarkable insight, into pregnant situations and connections; if it did, we might well despair of progress in understanding. Perhaps the difference between the primitive and the highly civilized man lies less in the explanation than in the observed connection of events. Certainly very significant observations have been made by those whose thought took the form of myth. Who, for example, would deny the significance of the early Babylonian records of eclipses because of their religious or astrological interpretation of the phenomena? The main fault to be found with them is that they did not also note the relative positions of the earth, the sun, and the moon. In the field of psychology, where we have advanced but little beyond the earliest Greeks, so far at least as concerns "explanation," the mythological account may suffer little by comparison with the most modern, e.g., in the portrayal of the sudden coming to manhood of Telemachus under the inspiration of Athena.[6]

[6] Homer, *Odyssey*, I. 113—III. 384.

There is no need of our going in detail into Greek mythology in order to show how it anticipated at many points the reasoned conclusions of scientists and philosophers; but it is proper to remark that poetic cosmogonies evidently existed long before we have record of cosmologies, and that the Greek philosophers themselves recognized a connection between the two classes of accounts of the origin of the world. "Even a lover of myth," says Aristotle,[7] "is in a sense a philosopher;" and Strabo says,[8] "the first historians and philosophers of nature were writers of myths." Aristotle, again,[9] calls Thales the founder of the school of philosophy which inquires into the material cause of things; but he adds, almost in the same breath, "some think that the ancients, who lived long before the present generation, and first framed accounts of the gods, held a similar view of nature." According to Simplicius,[10] Theophrastus thought that, though Thales was the first recorded philosopher of nature, he had many predecessors, whom he, however, far surpassed and overshadowed. In speaking of the sea and its origin, Aristotle[11] goes back to Hesiod's *Theogony*, and passes on to the Milesian and later scientists and philosophers without noting any essential difference in point of view; and late writers make no distinction whatever between the two classes of thinkers, as when Hippolytus says,[12] "The poet Hesiod himself declares that he thus heard the Muses speak about the philosophy of nature." Plato, playfully turning the tables on the philosophers, says,[13] "Each appears to me to recount a myth for the sake

[7] *Metaph.*, I. 2, 982 b 18.
[8] I. 2, 8.
[9] *Metaph.*, I. 3, 983 b 20 ff.
[10] *Physic.*, p. 29, 23 ff., ed. Diels.
[11] *Meteor.*, II. 1, 353 b 29 ff.
[12] *Philosophumena*, 26.
[13] *Sophist*, 242 c.

of entertainment, as if we were children.   One says that all things existing are three in number, and that certain of these somehow go to war with each other from time to time; then again they become reconciled, contract marriages, beget children, and rear their offspring.   Another says there is a pair—Moist and Dry, or Hot and Cold—and gives away the bride and lets the pair cohabit.   The Eleatic tribe out our way, however, going back to Xenophanes and even farther, recounts its tales as if all beings, so called, were one." It is obvious that Plato notes and emphasizes the fundamental identity in point of view between the early cosmogonists and the golden tribe of philosophers, and he shows how easy it is to state philosophical conceptions in mythological terms, implying that the opposite procedure is equally easy.

Aristotle also clearly correlates "theologians" and "theology" with "physiologists" and "physiology" in such sort as to show that in his view words and concepts are parallel.[14] He compares the earliest science to a lisping child, and makes repeated attempts to restate in more acceptable terms the opinions of his predecessors, as scientists are continually doing to this day.   He would doubtless have offered a like apology, only with larger charity, for the still earlier cosmogonists.   In the same spirit Theophrastus[15] remarked upon the "poetic" diction of Anaximander, because he spoke of the mutual encroachment of the elements as acts of "injustice."   In fact, the mythical cast of much of the earlier science and philosophy is so marked as to present a serious problem to the historical student who would fairly interpret the thought of the age.   Light is thus thrown in both directions.   One sees that philosophical, or quasi-philosophical, notions are clothed by the theogonists and cosmogonists in a garb of myth, and that the Greek

[14] E. g., Metaph., X, 5, 1071 b 26 ff.
[15] Simplicius, Physic., p. 24, 20, ed. Diels.

philosophers were not blind to the fact. While the former pictured the origin and operations of the world as the history and behavior of mythical characters, often so vaguely and imperfectly conceived as at first sight to betray their fictitious nature, philosophers, perhaps not so conscious of what they were doing, applied to their principles and elements names and epithets proper to the gods. This course was indeed extraordinarily easy and natural to the Greeks, whose religion was in its higher phase in great part a worship of Nature. Conversely, this worship of Nature in her more significant aspects predisposed the Greeks to a reasoned study of nature, once their thought became more truly reflective.

We shall presently have to consider more in detail certain corollaries of the relation of philosophy and science to theogony and cosmogony. That it is of great importance is certain; but there is, of course, a difference between them—a difference the Greeks were aware of even when they emphasized the continuity of the tradition. The important question is just wherein they thought the difference lay. Though, as we have remarked, Aristotle reckoned science and philosophy among the employments of leisure for pastime, they were not regarded really as forms of play. Even poetry among the Greeks was commonly regarded as having a serious purpose. Horace not inaptly expressed the Greek view when he asked, *ridentem dicere verum quid vetat?* Though the epic poet might be called an idle singer of an idle day, Homer was esteemed a teacher, and became perhaps the most important text of the Greek school-boy. The heroic epic was, however, succeeded by the didactic, and Hesiod represents the Muses as saying to him, "We know to speak full many things that wear the guise of truth, and know also when we will to utter truth."[16] The significant

[16] *Theogony*, 27 f., tr. Mair.

change came with the substitution of prose for verse, which was generally, though of course not always, likewise a change from myth to reasoned account. In time there came to be accepted a corresponding distinction between *mythos* and *logos*, the latter being the result of *historia*, study or research. Plutarch, explaining how the ancient custom of uttering the Delphic oracles in verse came to be abandoned in favor of prose, connects the change in practice with the general simplification of life and manners, and says, "History descended from her lofty meters, as from a chariot, and in pedestrian prose best gleaned truth from myth; and philosophy, esteeming clearness and intellectual enlightenment above astonishing effects, pursued her quest by reasoned argument."[17] Thus, though there was not wanting a distinction in substance, the difference was primarily one of form and spirit. No one, perhaps, has more clearly drawn the distinction between truth and fiction than Plato; the use which he makes of myths in his dialogues is therefore highly instructive. Without claiming that the statement exhausts the subject, one may fairly say that he employed them for ethical reasons, because the appeal to the imagination and the emotions might be counted on to reenforce the logical argument designed to convince the reason. The very fact that the Greeks recognized and consciously utilized this distinction may be reckoned among the best evidences that they had attained a clear conception of science.

The questions asked by science and philosophy are the same as those which man has presumably always raised, the questions still asked by the normal child. As Sully says of the questioning of children, "From the first the 'why' and its congeners have reference to the causal idea, to something which has brought the new and strange thing into existence and made it what it is. In truth this reference to origin, to

[17] *Moralia*, p. 406 e.

bringing about or making, is exceedingly prominent in childrens' questionings. Nothing is more interesting to a child than the production of things. What hours and hours does he not spend in wondering how the pebbles, the stars, the birds, the babies are made. The vivid interest in production is to a considerable extent practical. It is one of the great joys of children to be able themselves to make things, and this desire to make things, which is probably at first quite immense, and befitting rather a god than a feeble manikin of three years, naturally leads on to inquiry into the mode of producing. Yet from the earliest a true speculative interest blends with this practical instinct. Children are in the complete sense little philosophers, if philosophy, as the ancients said, consists in knowing the causes of things. The discovery of the cause is the completed process of assimilation, of the reference of the particular to a general rule or law."[18] True and significant as this statement is, it calls for a supplement. Children ask not only "why," in its various meanings, including the purpose, but also "what" a thing is. The answer here also supplies the reference of the particular to the general, but in this case the general is a class instead of a law.

However widely the "why" and the "what" may seem at first blush to differ, the answers, among the Greeks at least, are ultimately much the same. To inquire what a thing is, calls for a definition; and a definition gives the genus and the specific differentiae, as one identifies a person by giving his individual and his family name. The Greeks asked, Who and of what parentage are you? The answer supplied the father's name (and officially, at Athens, one's local habitation) in addition to the given name. If the individual was of great consequence his entire lineage might be carried

[18] *Studies of Childhood*, 78 f.

back on his father's side[19] to a god as the ultimate author of
his being.   The process of tracing one's pedigree was known
as *genealogy*, and this term was frequently used in the sense
of *aetiology*, "determining the cause."   Thus identification
ultimately resolved itself into the task of finding the cause,
as the source of being.   One readily sees how this procedure
is related to logical definition and the family-tree of classification.[20]   The system of definition and classification, however, leads directly back to theogony.   We have only to
cast a glance at the first efforts of the Greeks to bring order
out of chaos by the systematization of events, which we
call history.   In every attempt the mind passes from the
given, the known, to the unknown.   In this matter, the
given was the present; and the task was to work back
to the beginning.   The indispensable frame was a time-
scale, the raw materials for which lay ready to hand
in the genealogies of the noble families.   Relatively fixed
points were to be found in certain events which could be
determined by several mutually supporting or correcting
genealogies; such events being the migrations or coloniza-
tions, the Trojan War, and the Argonautic Expedition.
Since the last-named enterprise was by several concurrent
traditions fixed at two generations before the Trojan War,
and since the Heroes who took part in it were sons of gods,
we have been brought back to the end of the theogonies.
Of the process by which the Greeks of the Sixth Century
created an absolute time-scale for universal history we need
not speak at present, although it is one of the greatest
achievements of the Greek mind in its effort to make its
world intelligible.   What calls for remark is the fact that
this process—an inductive one—has to be inferred and reconstructed, as we shall have to infer and reconstruct the

[19] Hence the Greek use of *patria*, in the sense of "pedigree."
[20] Cf. Plato, *Sophist*, 268 c–d.

actual mental processes of the early Greek scientists and philosophers, while the order in which our sources present the matter is just the reverse, a deductive process. As we find it in Hesiod the tale begins with Chaos and Black Night, passes through a list of largely fictitious divinities to gods who represent in general something actually in relation to the life the Greeks then knew. So far the *Theogony*. To bridge the gap between the generations of the gods and the historical present, created by the migrations and the colonization of the chief cities, the Hesiodic school furnished the *Catalogues*, which related the birth of the Hero-founders. The actual reconstruction of the past, of course, proceeded from the known present to the inferred past, while the order of presentation is the reverse—it is *history*.

As has already been suggested, the answer to the question "why" takes very much the same form, in so far as it looks to the cause or source of being. The human mind seems instinctively to seek the cause in an agent; certainly, the first cause was an agent. The hierarchy of causes might then be represented as a theogony, when man began to reflect and to systematize his ideas of the world. This was the more natural to the Greeks because one of their fundamental beliefs was that nothing was created *e nihilo* and because the primitive Greeks, like other primitive peoples, conceived of creation as procreation.[21] It is not necessary to dwell on the fact beyond calling attention to certain phrases in which this ancient notion perpetuated itself. Not to press such expressions as the "genesis" of the world and the "cycle of birth," in the sense of the cosmic process, it suffices to recall that Aristotle repeatedly speaks of "generating a cosmos" when he means no more than determining the process by which the world came into being, or giving a philosophical account of its origin; and derivative forms of

[21] Cf. my book, *The Day of Yahweh*, 413 f.

existence are called the "offspring" of the elements.[22]   One understands, accordingly, how the "elements" and the "first causes," even when consciously proposed as physical and mechanical antecedents of concrete forms of existence, were endowed with attributes which assimilate them to the gods; for, as Rohde well said, "whoever among the Greeks said immortal, meant god."[23]

We shall presently have occasion to speak of the two kinds of definition devised by the Greeks, the genetic and the teleological, and to note their difference; but it is well here to call attention to the fact that they have much in common, since, as Aristotle did not fail to perceive, origin and essence are but two aspects of causation.   That the two-fold aspect of "nature" consequent on the common point of view, which regarded existence as a process, led to formulas showing little logical difference, is not surprising.

Were it not for the danger of being misunderstood, one could not do better than to speak of Greek science as "natural history;" for, at least in its origin, it was meant to be a history of nature or the world.   It is even now one of the best ways of explaining the nature of a thing to describe its origin.   That is to give a genetic definition of it.   Any one who has dealt with children well knows how effective this method is.   Students of folk-lore are familiar with the supposed magical efficacy of the story which recounts the origin of anything desired: if the savage wants rain, he may either produce a mimic rain-fall or tell how rain was originally produced.[24]   It is presumably this age-old belief that

[22] Cf. Bonitz, *Index Aristotelicus*, 150 a 6 ff., 406 a 10 ff.; Xenophanes, fr. 33, Diels; Parmenides, fr. 10, 3, Diels.

[23] Immortality, to the Greeks, did not necessarily involve eternity. Though Aristotle regarded the world as a god and uncreated, the traditional view assumed the birth of the gods and of the cosmos.

[24] Cf. my Περὶ Φύσεως. *A Study of the Conception of Nature among the Pre-Socratics* (*Proceedings of the American Academy of Arts and Sciences*, vol. XLV, 1910, pp. 86 ff.).

accounts for the fact that the story of creation was part of the Babylonian New Year's rite: for at the turn of the year the world was supposed to be in danger of coming to an end —to recite the story of creation (a magical formula or recipe) was to re-create the world and insure its continuance.[25] Originally, then, the genetic definition had a practical as well as a speculative interest. Of course, it does even yet in many things: but we have in general learned to distinguish between sciences which have, or may have, a practical aim in production and sciences which are wholly or partially limited to the satisfaction of our desire for understanding. Failure to distinguish between them led in ancient times to vain hopes, as it has done in our own day. It is hardly necessary to cite ancient or modern instances.

Enough has been said by way of pointing out the original connection of Greek science with magic and religion, though much might be added in illustration of the influence exercised by points of view established in the dim past before the dawn of history. In order that our account may not be misleading we must now call attention to the effort of the Greeks to emancipate themselves from the ancient beliefs. That they succeeded in this attempt beyond any other people of the ancient world was perhaps due as much to good fortune as to their national character.

If the Greeks more than any other people of antiquity developed a sense of natural relations in every sphere of life and conduct the reason may in good part be found in the fortunate circumstance that their early history is that of small independent communities with a variety of cults, some of which they found existing in the land on their arrival, while they brought others with them. With the hospitality characteristic of primitive peoples, they adopted the old and allowed them to make such alliances as they could, or would, with the new. No constraint was used,

[25] Cf. *The Day of Yahweh*, 142 ff., 408 ff.

and therefore there was no official preference shown.   All
the sanctities were to be observed, but none could dominate
the rest.   Furthermore a cult seems to have belonged primarily to the family or the clan; as such it was restricted
and not disposed to impose itself on outsiders.   There were
priests, but priests only of local shrines.   In other lands,
where great monarchies were established, the priests of the
private cults of the kings might come in time to dominate
the land.   In Greece, there was nothing of this kind.   The
innate tendency to achieve liberty of action, which the
Greeks display from the earliest times, cooperated, no doubt,
to bring about the state of affairs which characterizes Greek
civilization at its height.   The state, whatever its religious
sanctions, was an essentially secular institution.   It was of
course not irreligious, nor unmindful of the due observance
of religious rites; but these were for the most part left to the
family or clan and its sacerdotal representatives.   Even
when a cult became, so to speak, official or established as
belonging to the state itself, the priests who were charged
with the due observance of the cultus were virtually or actually secular functionaries.   Thus institutional religion in
Greece, so far as the state concerned itself with it, was secu-
larized and did not develop a hierarchy.   Furthermore, institutional religion was purely ritualistic and did not concern itself with the interpretation of its rites.   The state
might, of course, take note of non-observance, neglect, or
profanation of rites, but what the individual might think
about them was for him to determine.   Only in case his
actions or influence might lead to the neglect of religion did
the state take cognizance of them.   There are extremely interesting instances, however, in which Greek states inter-
vened to forbid superstitious practices, even when they
touched matters as sacred as rites to the dead.[26]

[26] Cf., for example, an inscription of Iulis on Ceos, dating from the last
quarter of the Fifth Century B.C. (*I. G.*, XII, v, i, 593).

Gibbon's saying,[27] "Freedom is the first step to curiosity and knowledge," nowhere finds fuller exemplification than in the history of Greek thought; and nowhere did thinkers so much feel the need of asserting their freedom as in matters where there was an actual or possible clash with theology. There was, to be sure, no official dogma, as has been pointed out; but there did exist what might fairly be called an accepted theology, though it possessed only such authority as belongs to all beliefs that have become traditional. These beliefs were embodied in the myths and the formulations of the poets, which had become the common property of the people. It is hardly necessary to state that the philosophers had from the first broken with these beliefs at least in intention, however much they may have been haunted by presuppositions involved in them; for the very attempt to offer a rational account of things implies a dissatisfaction with previously accepted opinions. The Hippocratic *Law* puts the matter succinctly: "Science and opinion are two distinct things; the former leads to knowledge, the latter to ignorance."[28]

It would be a mistake to regard Greek science as irreligious, as some have done. There was, to be sure, a great deal of criticism of popular religious conceptions from quite early times, but with few and unimportant exceptions the great Greek thinkers were reverent and in the true sense religious men, if we may judge them by their writings. Nor was their attitude one of mere lip-service. It was a common saying of the ancients, and it is worth repeating, that philosophy, like science, originated in the desire to rid the world of confusion. They rest ultimately on the assumption of a certain fundamental unity in things, perhaps, in strictness, a moral postulate, which as it comes progressively more clearly to consciousness ramifies in many directions and

[27] *Decline and Fall*, ch. LXVI.
[28] Hippocrates, *Lex* (IV. 642 L.).

constitutes the frame that supports the entire structure of man's making.   Religion, of course, felt that need as well as did other forms of the higher life, though, because of the especial sanctity of the objects within its domain, it has always been most conservative.   Therefore the critics of religion, from Xenophanes to the latest, were really reformers who attacked what seemed to them false in the interest of what they regarded as true religion.   How this problem was faced by men of science can nowhere be seen more clearly than in the writings of the Greek medical men. Some modern scholars have misconceived the situation because they have too much emphasized particular words. There are a few passages in the Hippocratic Corpus which seem perfunctory, such as one might perhaps expect to find in an unbelieving modern who wished to avoid offending sensibilities he respected but did not share; as when one writer curtly remarks[29] that the physician, in making his prognosis, should take account of a possible divine intervention, and another says[30] that in matters human the divine is the chief cause, thereafter the constitutions and the complexions of women, and immediately proceeds to a detailed consideration of the latter, dismissing the supernatural.   In reality there is nothing questionable about this procedure, certainly not from the point of view of religion; for what more could a physician have said?   The characteristic attitude of the medical men, however, may be assumed to have been that of the author of the treatise *On Epilepsy.*[31]   When one considers the hold which magic and demonology had on the minds of men throughout the world in ancient times, and still has among the vulgar, es-

[29] Hippocrates, *Prognost.*, 1 (II. 112 L.); but cf. *De Habitu Decenti*, 6 (IX. 232 L.).

[30] Hippocrates, *De Natura Muliebri*, 1 (VII. 312 L.).

[31] Hippocrates, *De Morbo Sacro*, 1 (VI. 352 ff. L.).

pecially in medical matters, one can not sufficiently admire the enlightened Greek thinkers who combated these superstitious beliefs and practices.   Mental diseases were almost universally thought to be due to possession by spirits, and those who were afflicted with them were accordingly treated with especial consideration—one of the few commendable results of superstition.   Epilepsy in particular, because of the phenomena attending the spasms to which its victims are subject, was believed to be a divine affliction and was called "the sacred disease."   The author says frankly that in his opinion those who first so regarded epilepsy were like the impostors, quacks, purifiers, and "pious frauds" of his own time, who profess uncommon piety and hide their own want of resources behind an appeal to divine agency, lest their ignorance stand revealed.   Epilepsy is of divine origin, as is every other disease, no more and no less; for all have their basis and causes in nature.   The resort to magical practices amounts to impiety, because instead of recognizing the divine agency as supreme, it assumes that it is subject to human control.   As a matter of fact, the author assures us, epilepsy is due to decay of the brain—it is not confined to man, but occurs also among sheep and goats; and if one examines the brain of a sturdied goat, one discovers that it is filled with a liquid and has an offensive odor.

All diseases are divine and all human; but each has a character and cause peculiarly its own.   In another Hippocratic treatise,[32] possibly by the same author, we are told that the Scythians worship eunuchs because they attribute their condition to a god and fear a like fate for themselves. The writer adds, "I myself regard this affliction as divine, as well as every other.   One is not more divine or human than another; all are on the same level, and all are divine,

[32] *De Aëre, Aquis, Locis,* 22 (II. 76 ff. L.).

yet each has a natural cause, and none occurs without a natural cause." In the same spirit Lucian writes in a treatise *On the Syrian Goddess*,[33] in which he imitates the language, style, and mode of thought of the early Ionians. He tells the story of the Adonis River, giving first the mythical version and then the naturalistic explanation which in the manner of Hecataeus and Herodotus he attributes to a native of Byblus, and ends by saying, "Even granting that my informant gave the correct account, I think the coincidence of the wind quite divine."[34] Yet, however true to belief in the gods, the Greek man of science maintains a common-sense and practical attitude to life. In the Hippocratic Corpus there is a discussion[35] of dreams as possible aids to diagnosis and prognosis. The writer, after explaining that in sleep the soul retires from the members of the body, which therefore loses sensibility, distinguishes between dreams sent by the gods for the guidance of states or individuals and such as arise from repletion or lack of nourishment. There are persons, he says, who have developed an art of interpreting both kinds of dreams, but as for the latter kind they sometimes hit the mark and sometimes fail, because they know neither the causes of the dreams nor the reasons for their own success or failure. "They exhort one to take care lest one suffer some misfortune, without saying how one is to take care; they only bid one to pray (or make vows) to the gods. Prayer is, of course, good and proper; but one must help oneself as well as call upon the gods."

[33] *De Dea Syra*, 8.

[34] One may compare this attitude to that of Plutarch, *Vita Periclis*, 6, where he tells of the goat with the malformed head, which led to the prophecy of a seer and a dissection by the philosopher Anaxagoras. Both seer and philosopher, he suggests, were right, though from different points of view.

[35] *De Victu*, IV. 87 (VI. 640 f. L.).

One can not help being impressed by the sanity of these utterances; but they display no open hostility to religion. What one sees is rather the emergence of a conviction that science is concerned with proximate causes, and an emphasis on a new conception of the world which is summed up in the word Nature. This conception in the Fifth Century B.C. dominates Greek thought in every sphere. Without dethroning the gods, it did tend to eliminate them as proximate causes and by removing them virtually to an extra-mundane sphere to assign to them, so far as philosophy dealt with the question, the sole rôle of first cause in the physical world. In the moral world their reign was threatened by the growth of an ideal, which could ill brook the confusion and immorality of the ancient myths, and could find a place for piety only by unifying and purifying the old pantheon and by recognizing the new heaven thus constituted as a reflection of itself. For the time being the gods seemed to be virtually excluded from the world of Nature; so far as they received a place in it, they found it under the shadow of Law or Custom (*nomos*), which was generally opposed to Nature, but was sometimes regarded as its creature. We are not now concerned to follow the fortunes of the gods in Greek thought, but we must consider more at length the conceptions of Nature and Law.

There were, of course, even in the Fifth and Fourth Centuries B.C., intellectual leaders who attributed the world-order and the creation of individual things to the gods;[36] but Plato truly said[37] that the majority took the other view, which we find briefly expressed by Plutarch: "All becoming has two causes, of which the most ancient theologians and poets chose to turn their attention to the stronger only, pronouncing over all things the universal refrain: 'Zeus

[36] Hippocrates, *De Victu*, I. 11 (VI. 486 L.).
[37] *Sophist*, 265 c.

first, Zeus middle, all things are of Zeus,' while they never approached the necessary or physical causes. Their successors, called *Physikoi*, did the very reverse; they strayed away from that beautiful and divine principle, and refer everything to bodies, and impacts, and changes, and combinations."[38]

Aristotle repeatedly speaks of the earliest scientists as *physikoi* or *physiologoi*—as those who concerned themselves with *physis*. It is not necessary here to consider in detail the many senses in which that important word was used;[39] it may suffice to say that by the end of the Fifth Century it had developed practically all the meanings which we now attach to Nature. Perhaps the most important thing to note is that, possibly owing to the engrossing attention bestowed on the subject, Nature came to be regarded not only as a process in all its phases, but—what is most significant—as a process complete, self-contained, and all-inclusive. The logical consequences of this view were of course not at once contemplated or drawn; but, however little they came clearly to consciousness, they inevitably led in the end to a conception of the world which, as Plutarch says, left no place for the divine principle, except as Nature herself, by acquiring the functions formerly assigned to the gods, became in all but name a divine being. This consummation, however, was late in being attained, coming at the end of Greek history. Therewith philosophy had fetched full circle and returned to its point of departure. In the meantime, however, theology and science had to make such compromises as they could.

As for *nomos* (law), it seems first to have come to consciousness in political, social, and moral relations. Law appears always to begin as customary procedure, to which society becomes addicted as the individual does to habit.

[38] *De Defectu Orac.*, 436 d, tr. Prickard, modified.
[39] Cf. my Περὶ Φύσεως.

In a stable, homogeneous group, isolated from others, old customs become sacrosanct and are almost self-enforcing; only when the circle is enlarged to admit other groups, or when individuals pass freely to alien groups, and note or adopt other customs, will conflict arise and will established procedure require to be enforced by other sanctions.   Custom then becomes Law.   This stage was reached in Greece as a consequence of expansion resulting from commerce and colonization, and was marked by the first written codes.   These same social changes, by enlarging the intellectual horizon, eventuated in a quickening of the native curiosity of the Greeks, which led to conscious and critical appraisal of custom and law, as they existed in Greek lands, in comparison with those to be observed in the foreign parts made known by travel and report.   We need not follow the consequences in social life, except to point out that custom and law, now called in question, came to be regarded as opposed to Nature, whose norm was sought among primitive men and animals, probably because they were supposed to be guided by natural instinct (*physis*).   There were, however, far-seeing men who perceived that ultimately the habits, which result in custom and law, spring from Nature, so that at best the distinction between Nature and Law is one of degrees—degrees of immediacy.

But law (*nomos*) outgrew its relation to social life and was referred to the procedure of external nature.   It is interesting and characteristic of Greek thought, however, that the unquestioned instances of the use of *nomos* in the sense of a "law of nature" in the physical world are quite rare, while in the moral sphere "laws of nature" and the "laws of God" are freely contrasted with man-made statutes or decrees; for to the Greeks *nomos* was essentially an order embodying a command and supported by authority and sanction.   That the analogy of civil order was not absent from their thinking is evident from the way the Greeks had of speaking of cosmic justice (*diké*); but the terms most commonly employed with reference to what are now generally miscalled natural laws are "necessity" (*ananké*), "principle" or "formula" (*logos*), and natural process (*physis*).[40]   The Greeks were keenly aware of the existence of established order in the world, to which they gave the name of Cosmos; but since that order was regarded as autonomous, the expression of a process complete and self-contained, and not something impressed upon it from without, they seem to have avoided the term "law" or, when they used it, apologized for it as for a dubious metaphor.   We must never forget that science has from the first concerned itself with nature in two fields, which have to this day remained distinct despite the most determined efforts to merge them in one—the field occupied by living and thinking beings and that which includes "dead matter."[41]   In the latter it has been possible to state the results of investigation in mathematical-mechanical terms with ever-increasing success, whereas in the former, though physiology has in part become a branch of chemistry, psychology and ethics, not to speak of other disciplines, have proved intractable.   A leading biologist has lately confessed that in his field it is impossible to get on without teleology; and one sympathizes with the logical procedure of the Socratics in defining things with reference to the ends they subserve[42] when one contrasts it with the genetic method which obtains in chemistry.   Some years ago a chemist stated that one could make very good

---

[40] Cf. Rudolf Hirzel, *Themis, Dike, und Verwandtes* (1907).

[41] Cf. Kant, *Träume eines Geistersehers*, ch. II, *initio*.

[42] Cf. Hippocrates, *De Alimento*, 21 (IX. 104 f. L.): "Nutriment is not nutriment, if it can not nourish; non-nutriment is nutriment, if it can:— nutriment in name, but not in fact; nutriment in fact, but not in name." This is a clear case of definition by function.

milk synthetically in the laboratory; meaning, of course, that the product when analyzed yielded the same formula as that obtained from milk.   When asked whether the synthetic milk could be digested and used for food, he innocently replied, "No, not that!"   The two distinct fields of scientific endeavor gave rise to distinct logical methods, as will later be pointed out at greater length.   For the moment, we are concerned only to call attention to the fact that, while Greek thinkers were alive to the existence of orderly processes throughout nature, they did not fail to recognize the difference between the fields above mentioned, and with the fine discrimination characteristic of all their thought and speech, used distinct terms to denote the uniformities which they observed.

The early Greek thinkers might have said, with the religious reformer, "the world is my parish;" for, at least in intention, their thought went out to the whole of nature.   We have been the victims of preconceptions, partly our own and partly inherited from those Greeks who first undertook to write the history of the sciences and of philosophy.   The categories which we use in the study of Greek thought are for the most part later in origin than the things to which we apply them; perhaps medicine is the only art or science which may with some propriety be said to have had a separate existence at the point of time where we date the beginning of reflective thinking about nature.   Even there, however, one may confidently say, the nomenclature is convenient rather than exact; for, on the one hand, the devotees of medicine were certainly concerned with all nature, so far as they knew it; and, on the other hand, every one who took nature for the object of his inquiries was bound to consider matters which in our view belong to medicine.   One may say that at first science was one, and a man either was or was not a scientist.   Nevertheless the

practical necessities inevitably led men to give more attention to subjects germane to the art of healing; and from the
first the relations between "physicians" and "philosophers"
were of the closest.[43]   One would gladly know far more than
the scanty record vouchsafes to us about the early voyages
of discovery undertaken at the behest of Darius Hystaspes,
by which a first acquaintance was made with the Caspian
Sea and with India.   In both cases the fragmentary reports
disclose that besides the all-important geographical information, data concerned with botany were collected.[44]   We
know that Greek physicians from the most ancient times
derived much of their materia medica from very distant
lands, doubtless for the most part in the way of overland
traffic along immemorial trade-routes.   Articles so brought
were generally small, such as simples, and the prices they
commanded would pay the freight.   Since trade was presumably largely in such merchandise, and the enlightened
king was bent on increasing his revenues as well as his
power, one might guess that he placed his expeditions under
the charge of men of science.   It is therefore of unusual
interest to know that the one which went to India, and
probably descended the Ganges, was entrusted to Scylax of
Caryanda,[45] a Carian town only a few miles from Cnidus;
and one has good reason to guess that this man was a Greek
and that he stood in rather close relation to the medical
schools of Cnidus and Cos.   This is the more probable because we learn that Darius conceived an expedition of exploration as a preliminary to conquest in the West, and ap-
pointed as its leader his Greek body-physician Democedes,[46]

[43] Aristotle, *De Sensu*, 1, 436 a 19 ff.; *De Part. Animal.*, I. 1, 640 b 4–17.

[44] Hecataeus of Miletus, fr. 172 ff. (Müller, *Historicorum Graecorum Fragmenta*, I. 12).

[45] Herodotus, IV. 44.

[46] Herodotus, III. 125, 129–137, tells his story.   Compare the story of Eudoxus of Cyzicus, told by Posidonius (Strabo, II. 3, 4), and that of Ctesias, told by Diodorus Siculus, II. 32, 4.

who came from another great medical center, Croton in Sicily, an off-shoot of the school of Cnidus.   These facts help us to understand the beginnings of science, for its devotees were strictly naturalists in the widest sense.

The fragmentary record of the early times is calculated to mislead us; but even more, our way of approaching the subject.   If one examines the older works on the history of philosophy one finds that they are, like a recent popular book, essentially biographical, spiced with the chit-chat of old Diogenes Laertius.   Since the actual scientific or philosophical data are so few, and so scattered, one naturally concludes that the early thinkers were very casual in their method and had little in common.   A closer scrutiny will, however, go far to correct that impression, for which the earliest systematic account of philosophy is responsible. This was the treatise *On the Opinions of the Natural Philosophers*[47] written by Theophrastus, the pupil and successor of Aristotle in the Lyceum.   To understand this work and the misconceptions to which it has given rise, it is necessary to consider its form and its place in a larger scheme.   Aristotle, whose interests and writings were encyclopedic, had followed the plan of introducing the discussion of special topics with a brief survey of previous thought on the subject. How he collected the data for these summaries we do not know, though it is not unlikely that he was aided by his associates, who were naturally dominated by his own points of view, as pupils to-day are influenced by the doctrines of their teachers.   However biased he may have been in any given case, Aristotle, though not by nature historically minded, was sincerely desirous of promoting historical studies in the sciences; and consequently a more or less extensive program was evolved and carried out by his associates.   Theophrastus prosecuted the study of botany and Natural Philosophy;

[47] The fragments and derivatives were collected and edited by Hermann Diels, *Doxographi Graeci*, Berlin, 1879.

Eudemus took astronomy and mathematics for his field; and Meno wrote the history of medicine.   The vast subject of the constitutional history of the almost innumerable city-states of Greece must have called for the cooperation of the whole force of investigators.   Furthermore, the men above mentioned, and others whom we pass over, followed their own predilections in cultivating other fields.   One is filled with admiration as one contemplates the scope of this enterprise in historical research, to which we owe an immeasurable debt of gratitude.[48]

Of the shortcomings of these historical surveys, we need not speak at present; but several points call for remark. First of all we must note the consequences of the division of labor between a number of workers.   When Eudemus and Meno treated separately of the mathematical sciences and medicine, these fields were definitely set apart from the domain of Natural Philosophy; consequently the doxographic tradition, which derives from Theophrastus, ignores data proper to them, and modern historians of philosophy have until quite recently paid little attention to such matters, even when they were known to have been dealt with in works supposed to be philosophical.   The Aristotelian conception of philosophy has in fact entirely dominated the historical study of the subject; and the earliest Greek thinkers appear to have found a place in the survey solely because Aristotle thought he recognized in them a preparatory stage to his own philosophy.   One important consequence of the division of labor, then, was the total omission, or the comparative neglect, of matters which deeply concern science and philosophy, and which we know from other sources engaged the serious attention of the early scientists.   Not to speak of the subjects treated by

[48] Cf. Usener, "Organisation der Wissenschaftlichen Arbeit," *Vorträge und Aufsätze*, 69–102.

Eudemus and Meno, of whose treatises very meager fragments survive, one may instance the sad state of our knowledge regarding so important a subject as physical and descriptive geography. We know, to be sure, that the early
Milesians Anaximander and Hecataeus were deeply interested in this subject and made very important contributions
to the scientific study of it, but Hecataeus, the most eminent
geographer before Eratosthenes, was not even mentioned in
the doxographic tradition.[49] We should really know nothing of consequence about the first stages of inquiry in this
field if we did not have a quite distinct line of tradition—
the historico-geographic—to draw on. This study grew
up and flourished without the benefit of philosophers, although it touched subjects with which they had to concern
themselves. Plato displayed little interest in any sciences
except the mathematical, though he touched medical matters occasionally, particularly in the *Timaeus*, and geographical speculations in the *Critias*. Though acquainted with
the historical literature, he used his knowledge chiefly for
purposes of illustration, except in the *Republic*, *Politicus*,
and *Laws*, where his historical studies color his speculations.
Aristotle, as the son of a physician, was at home in the medical literature; in history, except in special fields, he seems
to have taken about the same interest as Plato; in mathematics he was not an adept and did not really understand
the higher developments reached in his time; and of geog-

[49] In his *Doxographi Graeci* Diels referred the statement about the sun
(p. 351 b 9) to Hecataeus of Miletus; but he subsequently recognized, what is
obvious, that it belongs to Hecataeus of Abdera, who lived at the close of the
Fourth Century B.C. Eratosthenes is mentioned by the Doxographers, but
only for "philosophical" opinions. Whether the treatise *On the Inundation of
the Nile*, ascribed to Aristotle by tradition, belonged to him or to Theophrastus, is uncertain. One would more naturally think of Dicaearchus as
its probable author, because its sources belong to the historico-geographical
tradition, in which he was deeply interested.

raphy, as distinguished from cosmography, he gives evidence of no special knowledge, though there are indications in his *Meteorology* that he had a superficial acquaintance with current maps.   It is a pity that his pupil Dicaearchus, a man of true scientific spirit and intelligent interest in matters of history and geography, did not give us an account of the development of these disciplines.

There is another characteristic of the work of Theophrastus that has greatly influenced the traditional view of early Greek science.   As has already been stated, Aristotle liked to give a rapid sketch of the opinions of his predecessors on a particular subject as an introduction to his discussion of it.   It was evidently in these sketches that his successor Theophrastus found the model for his account; for he did not present the opinions of a man as a whole and indicate their relations to one another or to the facts upon which they were based; but, dividing the subject into topics, he first enumerated the doctrines which seemed to him important, and then subjected them to criticism from the Aristotelian point of view.   Had his work been preserved intact we should, therefore, have had what we now have in Aristotle's works, only in larger compass and in richer detail.   As it is, we have meager extracts from his treatise in a succession of later epitomes, which, however, in a measure compensate their omissions by the addition of details derived from sources, partly of later date than Theophrastus, partly belonging to fields which he did not include in his survey.   Such being the character of the systematic accounts dating from antiquity, it is not difficult to understand why, in the sad wreck of the original works, the impression one obtains from a superficial study should be not unlike that which a ruined city makes on a casual visitor who lacks either the time or the schooled imagination to reconstruct the whole.   If now we add the further consideration, which is too commonly overlooked, that the interest of Aristotle and Theophrastus, which determined the choice of topics as well as the light in which the historical data were placed, was predominantly metaphysical, we are prepared to see that the problem of seeing early Greek science whole and in its true relations is not easily solved.

Fortunately we are not, however, entirely destitute of indications regarding the way the student should take in this difficult task.   Though this does not purport to be a history, the reader clearly has a right to demand at least the general grounds for the faith which justifies one in undertaking even such an attempt as is here made to speak of Greek science as a whole.   The answer must for the present be given in general terms, because to attempt to prove the correctness of the writer's position by citation of chapter and verse would be out of place, though the conclusions are the result of many years devoted to the study of the problem.   Perhaps the most fundamental conviction of the writer may be summed up in the statement that Greek science was from the first one, and was divided only for convenience and out of deference to the limitations of the individual; a corollary, or consequence, of this basic truth is that, despite the superficial differences, there is a striking unity in the view of the world held by the Greeks from beginning to end.   That science always had in view the whole of Nature one can easily see from the range of topics to which even the lesser intellects addressed themselves, while in the extant works of the great thinkers the effort is unmistakable "im Ganzen, Guten, Schönen resolut zu leben."   The fundamental unity of their thought, to recognize which requires, to be sure, more than superficial acquaintance, Usener would perhaps have attributed to the circumstance that their stock of ideas was bred in the night preceding the dawn of history, when unconsciously or subconsciously the character of the race was shaping its destiny.   Besides, even from the period before Plato and Aristotle, we possess a considerable body of literature consisting of whole treatises, in which one may see the working of the Greek mind.   Among these we may single out as significant for our present purpose the history of Herodotus and the repository of medical literature known as the Hippocratean Corpus.

In claiming for Greek thought a striking unity, one must not be understood as denying that it actually developed in many directions; on the contrary, one might with considerable show of reason assert that in its course it at least foreshadowed all the major forms which the thinking of the Western world has assumed.   This is so generally acknowledged that there is no need of insisting on it.   The reasons for the fact are perhaps not so apparent.   The direct dependence of modern thinkers on their ancient predecessors no doubt accounts for much, but there remains the further, and more important, question why the modern world has so largely fallen in with Greek views instead of rejecting them; for it is not a case of simple continuation.   It is true that in one significant regard the tradition has been virtually uninterrupted; for Christian theology, which took form under Greek influences, has remained essentially unchanged. But in the brooding night of the Dark Ages a new life, fathered by new-comers in the western world, was slowly incubating; when it saw the light it looked on things in a different way.   What really calls for explanation is the eagerness with which it took in and assimilated anew what the Middle Ages had largely lost.   The reason must be that the new world instinctively felt its kinship with the old. Of course the process of assimilation is still going on; for ancient thought seems to have a perennial vitality, the possibilities of which in giving rise to ever new developments are apparently inexhaustible.

Greek thought was characterized by a healthy union of
realism with idealism. It accepted the world, but with
reservations—that is to say, it never seriously doubted
the reality of things, but it felt from the first that they
called for scrutiny and interpretation. If in the beginning
the acceptance was most marked, in process of time interpretation gained in depth and prominence. The earliest
Greek thinkers lived on the plane of science, but gradually
the insistent demand for interpretation led their successors
to assumptions distinctly metaphysical. The way back
from the supersensuous world, which metaphysics posits, to
the things of practical experience is difficult—far more difficult than the flight from the temporal to the eternal.
What distinguishes Greek thought from any other that has
dared to soar so high is the conscious need of relating the
two worlds by finding in the entities realized by the reason
the effective explanation of the things of sense. With
the success or failure of the attempts, we are not now
concerned.

The reference to metaphysics calls for an additional statement. Ontological problems were of course raised as soon
as men became aware that things needed explanation—
were not precisely what they seemed; but the first answers
were by no means ontological, in the sense that what was
assumed by way of explanation was somewhat beyond the
reach of ordinary experience. Soon there were presented
theories which strained to the utmost the capacity of interpretation on the assumption of matter, as we know it by the
senses; but if their proponents were aware of it, they did not
betray themselves. This is notably true of Heraclitus and
Parmenides. Only when Socrates turned with conviction
from the study of the external world to the contemplation
of ideals, and recognized in them the things most important
and consonant with human nature, did metaphysical prob-

lems clamor for clear statement and solution.[50]   In a real
sense philosophy, as distinguished from science, begins there.
But the logic of Socrates's position, as well as his place in
the line of the Orphic-Pythagorean tradition, made him the
advocate and exponent of an ethico-religious ideal of life,
which thereafter appropriated the name of philosophy.
Even when science shared the hospitality of the philosophical schools and perhaps was combined with philosophy
in the same person, the two were distinct.   By the discriminating, the names "sage" and "philosopher" were restricted
in use to those who pursued the ethico-religious ideal of life;
the scientist was called by other names, depending on his
special interest.   Yet it was in the period which reserved its
most honorific titles for others that men of science made
their greatest contributions to knowledge.

Explanation to the Greek meant the search for the cause;
but the notion of a "cause" is notoriously difficult to define.
Before the time of Socrates, definition of notions was almost unknown, except in mathematics, where little help
could be found for such a concept.   One must bear this in
mind in considering the earliest Greeks and not expect too
much precision in their thinking.   To the uncritical early
thinkers it meant simply that which one alleged in explanation of a fact or an occurrence.   What men were endeavoring to do was to make the world, as they found it, intelligible to themselves.   When Aristotle came to consider their
procedure, he confronted what he knew of their thought
with his own scheme, which recognized four uses of cause:[51]
(1) the material (of which a thing is made), (2) the motor
(which sets going the process leading to its production), (3)
the formal or definitional (which states its essential character), and (4) the final (the purpose for which it exists or

[50] Cf. Aristotle, *Metaph.*, XI. 1, 1069a 25–30.
[51] *Ibid.*, I. 3, 983 a 24 ff.

comes into existence). This scheme he at times simplified by consolidating all but (1) the material in (4) the final cause, as involving the other two. Recognizing that his final cause was the logical outcome of the Socratic search for definitions in the realm of ethics,[52] he generally limited the earlier thinkers to the investigation of the material cause. He was then at once confronted by the fact that Empedocles, for example, did have something to say of motor-forces operating to make and unmake successive worlds, and in other ways he betrayed his realization that his analysis did not quite fit the facts, notably by hinting occasionally that the earlier thinkers did not comprehend the full import of their words.

Aristotle says,[53] "Now that with which the ancient writers, who first philosophized about Nature, busied themselves was the material principle and the material cause. They inquired what this is, and what its character; how the universe is generated out of it, and by what motor-influence —whether, for instance, by antagonism or friendship, whether by intelligence or spontaneous action, the material substratum being assumed to have certain inseparable properties; fire, for instance, to have a hot nature, earth a cold one; the former to be light, the latter heavy. For even the genesis of the universe is thus explained by them. After like fashion do they deal also with the development of plants and animals. They say, for instance, that the water contained in the body causes by its currents the formation of the stomach and the other receptacles of food or of excretion; and that the breath by its passage breaks open the outlets of the nostrils, air and water being the materials of which bodies are made; for all represent nature as composed of such or similar substances." Excepting the ref-

[52] *Ibid.*, I. 6, 987 b 1 ff.
[53] *De Part. Animal.*, I. 1, 640 b 4 ff., tr. Ogle.

erence to spontaneity, which is made from Aristotle's own point of view,[54] this statement may be accepted. In another passage Aristotle contrasts the attitude of the Socratics with that of the early Naturalists and says,[55] "Generation follows essence and is for its sake, and essence does not follow generation. The ancient Naturalists held the opposite opinion, because they did not see that there are more causes, but noted only the material and the motor without even distinguishing between these; they were unaware of the definitional and final causes." In other connections he carries the analysis farther and regards the "elements" as the material causes;[56] for the things of common experience are composite, and the simple ingredients ("elements") form them.[57] "Let us assume that an element is that into which other bodies are resolved . . . . but which can not itself be resolved into others generically different; such is the meaning which all men attach to an element in all relations."[58] The existence of such elements, he goes on to say, is proved by their being separated out—by analysis, we should say. In this connection he introduces the distinction between the actual and the potential presence of the element in the compound—a distinction which is peculiar to him and those who accepted his metaphysics. The early scientists thought of properties as qualitative material ingredients actually present in the constitution (*physis*) of things. Plato still speaks after the manner of the Pre-Socratics when he says that one who would give an acceptable account of nature must attend narrowly to the variety of things, which is due to the mixture of the ele-

---

[54] Contrast *De Part. Animal.*, I. 1, 641 b 20 ff. with *Physic.*, II. 4, 196 a 5 ff.
[55] *De Generat. Animal.*, V. 1, 778 b 5 ff.
[56] *Metaph.*, X. 1, 1069 a 18 ff.
[57] Cf. Plato, *Theaetetus*, 203 a ff., 206 e ff.
[58] Aristotle, *De Caelo*, III. 3, 302 a 15 ff.

ments.[59]   So the medical writers say "the constitution of the body is the starting-point of medical science;"[60] or again "I assert that he who would write a proper treatise about human diet must first know and distinguish the human constitution—know of what it (the generic constitution) was composed to begin with, and to distinguish (in the individual) by what components it is controlled; for if he does not know man's original composition, he will be unable to know the effects of the components, and if he does not distinguish the component that overmasters the others in the (particular) body, he will not be in position to offer the (food or medicine) suitable to the man."[61]

In order to understand a thing, one must analyze it into its simple elements, which can not be further analyzed, and these are its least parts.[62]   These originally constituted it, and into them it is finally resolved.[63]   The conception of an element, however, was not at first sharply defined.   Only gradually did Greek thinkers realize the full implications of "unity," because a unit was at first taken in a quite practical sense.   In mathematics, for example, number in general, and the unit in particular, was conceived as a solid (concrete number); indeed the clear distinction between the (solid) "one" and the (abstract) "unity" (*monad*) seems to date from about the time of Plato, or a little earlier. However, the Eleatic subtleties about the "One" and the "Many" and the problems of motion focussed the attention of all on the concept of unity; and it was only when the Eleatics identified the "one" with the "homogeneous" that

[59] *Timaeus*, 57 c f.

[60] Hippocrates, *De Locis in Homine*, 2 (VI. 278 L.); cf. *De Ventis*, 1 (VI. 90 f., L.).

[61] Hippocrates, *De Victu*, I. 2 (VI. 468 L.).

[62] Aristotle, *Metaph.*, I. 3, 983 a 24 ff.; *Polit.*, I. 1, 1252 a 18 ff.

[63] Hippocrates, *De Natura Hominis*, 3 (VI. 38 L.).

it became necessary to scrutinize the notion of an element and to draw the logical consequence that if properties actually are present and permanent in matter there must be a number of elementary substances. Since, as Aristotle repeatedly assures us,[64] the early Greeks rejected the idea of creation *e nihilo*, there were but three ways open to explain the apparent variety in the world: (1) qualitative differences might be regarded as original and unchangeable, inhering actually in forms of matter divisible into very small corpuscles; (2) they might be conceived as merely potentially present in a common material substratum, to be brought out by conditions; (3) they might be regarded as having no primary reality, but as being secondary consequences of the shapes or collocations of particles (atoms) possessing no intrinsic properties except the geometrical. The second of these alternatives was taken by Aristotle, the third by Plato and the Atomists. After Empedocles the "elements" were commonly limited to four—fire, air, water, and earth; and this rough classification was for practical (and cosmic) purposes adopted by thinkers who well knew that "earth," for example, was strictly speaking not an element at all, because it was not homogeneous. A great deal of confusion (and misplaced criticism) is due to this loose use of the word "element" for what Lucretius well called the *maxima mundi membra*.[65] At best, however, any analysis of the physical world into elements of which the Greeks, with their limited apparatus and undeveloped technique, were capable would have been from the modern point of view equally unsatisfactory: what should and does concern us at present is the recognition of the logical problem and the fact that an analysis, however rough and

[64] *Physic.*, I. 4, 187 a 26 ff.; *De Generat. et Corrupt.*, I. 3, 317 b 28 ff.   Cf. Lucretius, I. 150, etc.
[65] V. 380.

provisional, was attempted.   The really important thing in
science is the idea; once given, it must be left to work itself
out in practice.   With the refinements suggested by experience and made possible by the perfection of methods
used in the arts, the analysis of matter into elements has
been carried vastly further than any Greek could have
dreamed; to-day there are indications that the number of
true "elements," however defined, may be reduced—possibly to one!

William James said:[66] "Every scientific conception is in
the first instance a 'spontaneous variation' in some one's
brain.   For one that proves useful and applicable there are
a thousand that perish through their worthlessness.   Their
genesis is strictly akin to that of the flashes of poetry and
sallies of wit to which the instable brain-paths equally give
rise.   But whereas the poetry and wit (like the science of
ancients) are their 'own excuse for being,' and have to run
no further test, the 'scientific' conceptions must prove their
worth by being 'verified.'   This test, however, is the cause
of their preservation, not that of their production; and one
might as well account for the origin of Artemus Ward's
jokes by the 'cohesion' of subjects with predicates in proportion to the 'persistence of the outer relations' to which
they 'correspond' as to treat the genesis of scientific conceptions in the same ponderously unreal way."   The brilliant psychologist, unquestionably right in his psychology,
was unfortunately little at home in the real scientific literature of the ancients; had it been otherwise, he must have
seen that the fundamental ideas and methods of science
were largely the "spontaneous variations" of Greek brains:
the process of testing them is still going on, and in the process the ideas and methods then suggested, like those suggested and tested in modern times, undergo modification

[66] *The Principles of Psychology*, II. 636.

and re-definition.   We shall have to recur to this process in speaking of scientific method.   As for the originator of the conception of an "element," we can not name him; we can not even say how it was suggested, though one may easily see that it has obvious relations to the process of classification, which, in the concrete, is close to the foundations of intelligence.

There are other fundamental conceptions, of unknown origin, which were clearly apprehended by the Greeks and which underlie all developments of science.   Such is the principle that everything has a natural cause.[67]   It ought perhaps to be called a postulate, because it has not been proved, and can never be proved, except in special cases: nevertheless, though it really is only the express demand that nature shall be reasonable and intelligible, it is the prime mover in all scientific endeavor.   This principle takes various forms, such as the denial of chance,[68] and the assertion of the reign of law; but it leads in the end to the assertion that the relation between antecedent and consequent, between cause and effect, is constant and invariable—to the belief in the uniformity of nature.[69] There were, of course, lapses from this high faith, but on the whole the Greeks were singularly true to it.   Various guesses have been ventured by writers, ancient and modern, as to the phenomena which first attracted man's attention and impressed him with the regularity of nature.   Plato[70]

[67] Leucippus, fr. 2, Diels; Hippocrates, *De Arte*, 6 (VI. 10 L.); *De Aëre, Aquis, Locis*, 22 (II. 76 f., L.).

[68] Aristotle, *Physic.*, II. 4, 196 a 5 ff.

[69] Archilochus, fr. 74, Diehl; Herodotus, V. 92 a, VIII. 143; Sophocles, *Philoctetes*, 1329 f.; Euripides, *Medea*, 411 ff.; Horace, *C.*, I. 29, 10; Hippocrates, *De Septimestri Partu*, 9 (VII. 450 ff., L.), *De Natura Hominis*, 5 (VI. 42 L.), *De Carne*, 18 (VIII. 614 L.).

[70] *Timaeus*, 91 d ff.

and Aristotle[71] thought of the occurrences in the heavens; and it is certain that the regular succession of day and night, the orderly sequence of the moon's phases and of the seasons following the advance or retirement of the sun, vitally affected primitive man and predisposed him to believe in an ordered world. But, while there were thus given data seemingly calling for the recognition of natural causes, the mind was not prepared to take that step; it conceived of all occurrences as determined, like man's own, by will and by favor or disfavor, and hence contrived a magical means of controlling them. Perhaps the first great quasi-scientific hypothesis was that of astrology, which arose when the planets came to be more closely observed and their relations were studied in addition to sun and moon. Mill says,[72] "That every fact which begins to exist has a cause, and that this cause must be found in some fact or concourse of facts which immediately preceded the occurrence may be taken for certain. The whole of the present facts are the infallible result of all past facts, and more immediately of all the facts which existed at the moment previous. Here then is a great sequence, which we know to be uniform. If the whole prior state of the entire universe could again recur, it would again be followed by the present state."

With very slight modifications this statement might be accepted as the principle of astrology; it would need to be supplemented only with the declaration that the prime factor which determines the secondary sequences is the position of the stars.

It is an observation of no little interest, and perhaps of incalculable consequence to the modern world, that the system of astrology did not dominate Greek thought. Why this was so is one of those mysteries which time has left as if

[71] *Metaph.*, I. 2, 982 b 12 ff.
[72] *Logic*, Bk. III., Ch. III, §1.

for the express purpose of tantalizing the historian. The discerning reader will have detected in many a myth known to the Greeks, such as the myth of Prometheus and of the feast of Thyestes, unmistakable traces of beliefs which clearly belong to the theory of astrology; but it is significant that the specific application of the principles, which alone make the myths intelligible, is nowhere in evidence before the Fourth Century B. C.[73] By that time the groundwork of Greek science had been laid, and whatever Stoics or other Orientals might import in the way of definite astrological conceptions could have but a superficial effect. In the Orient astrology and its corollary fatalism are very ancient and have left a heritage the influence of which in the lives of the peoples has made them inaccessible to science as the West conceives it.[74]

That the Greeks demanded, and did their best to create, an orderly world is clear from every point of view; but by instinct they were not fatalists. To adopt that attitude requires a peculiar mentality, which is certainly not that of the ancient Greek. One may say that it requires a heroic faith—the Greek was rather a sceptic, requiring evidence of fact or unanswerable argument for whatever he should accept. Perhaps it demands a renunciation of faith in oneself; if so, the active, practical Greek could not forego the effort to determine his own course. But, with all his assurance in action after deliberation, the Greek had a wholesome recognition of his limitations, particularly as regards knowledge of fact, as is abundantly shown by his restless curiosity and search for information about every-

[73] See, however, Franz Boll, in *Neue Jahrbücher für das Klassische Altertum etc.*, XXI (1908), 119, and Franz Cumont, *ibid.*, XXVII (1911), 5.

[74] I know no better expression of the spirit of the Orient than the letter of Imaum Ali Zadi, addressed to Sir A. Layard, which James, *The Principles of Psychology*, II. 640 f., quotes from Layard's *Nineveh and Babylon*.

thing. It is not mere chance that scholars who devote themselves to Greek studies so readily become pioneers in the most various lines of research; they can not avoid doing so.   If they catch the spirit of the people with whom they concern themselves, they will share their restless inquiries and will nevertheless not go beyond the evidence they find. Herodotus gives us several good illustrations of this tendency.   He sometimes doubts a report, but he will not reject it off-hand; for he is sensible of the limits of his knowledge and recalls that he has seen many things which were to him antecedently improbable.   He even reflects that the infinite lapse of past time offers scope for the occurrence of almost anything.[75]   Epicurus availed himself of the same hospitable privilege of the infinitude of time and space to introduce the minimal swerve of the atoms in their fall which must be assumed in order to account for the origin of the world.[76]   The Greek philosopher has been blamed for having recourse to this hypothesis; but the origin of the species, as Darwin presented it, was based upon a long chance not essentially different, and even a modern logician[77] might be quoted in justification of the assumption: "When the law [of uniformity] is interpreted in the looser sense . . . . I think that Mill speaks with more hesitation than he need adopt.   So far from admitting the bare possibility of a breach of uniformity in this sense, I should think it not at all unlikely that in the endless stretches of time and space there may be developments in store which fully deserve the name."   It is perhaps rather more than possible that we have in the assumption of Epicurus a faded echo of a very ancient belief, which appears in many forms.   The Greeks, like many other ancient peoples, ex-

[75] Herodotus, IV. 195, V. 9.
[76] Lucretius, V. 187 ff.
[77] Venn, *The Principles of Empirical or Inductive Logic*, 136.

pressed it in their myths, as of Uranus, Cronus, and Zeus, who represent cosmic ages. The birth, or accession, of a new Æon, which marks a new era, was regularly preceded by a transition stage of utter confusion, as the introduction of the Julian calendar was preceded by the "year of confusion." Students of Jewish and Christian eschatology are familiar with the Saturnalian subversion of all order and the topsy-turvy of morals expected in the "last days" as signs of the immediate approach of the new age. Earlier Greek thinkers had imagined a like Chaos, either before the world as they knew it began, or in the heyday of its youth—before Nature steadied itself in dependable ways. Whether this notion originated in astrology or was merely adopted by it, we do not know; but it is a regular part of that system as it has come down to us.[78]

The Greeks were preeminently a reflective people; and reflection bespeaks a disposition not to take things simply for granted. The results of their reflection usually assumed the historical form, telling how a thing originated; of course they considered the antecedents of science. It is plain that they regarded science as merely one mode of intelligence. Aristotle[79] traces knowledge, which all men desire, back to sensations. These are common to all animals, but some have, while others lack, memory; many sense-impressions retained in memory beget in man experience, and knowledge and art result from experience. Art arises, when from many notions gained by experience one universal judgment about similar things is formed. Science, however, differs from art in that art is practical and merely acts on a recognized truth applicable to a given class of facts; whereas science, though it may not be efficient in practice, knows the

[78] Cf. my article, "Vergil's Messianic Expectations," *Amer. Jour. of Philology*, XLV. (1924), 205 ff., and *The Day of Yahweh, passim.*
[79] *Metaph.*, I. 1.

cause and has framed a theory.   It is not mere chance that
Aristotle in this connection draws most of his illustrations
from the field of medicine.   It is true that here and elsewhere in Aristotle and Plato we can not point to the literary
sources which suggested their reflections, but the extraordinary frequency of their appeals to medical practice in illustration of scientific procedures, taken in conjunction with
the wealth of data to be found in the extant writings of
early physicians and the evidence of the Hippocratean Corpus itself for the existence of an abundant medical literature of early date now lost, justifies the assumption that
medicine along with mathematics[80] afforded the best examples of science in the Fifth and Fourth Centuries B.C.
While there is much of doubtful value in what has been
preserved, we have every reason to lament the irreparable
loss of works which, taken collectively, must have been of
capital importance.

The reader will excuse this digression; for the fragments
of the earlier philosophers in general throw little light on
the matters with which we are here concerned.   Xenophanes says,[81] "The gods did not reveal all things to mortals
from the beginning; but by searching in time they discover
the better."   The thought that Time discloses and teaches
all things is a Greek commonplace;[82] and everywhere one
meets the conviction that knowledge, like every phase of
human life, is subject to development.   Herodotus, for example, expressing his belief that the rites performed in the
service of Dionysus were learned from the Egyptians, says,[83]

[80] Plato, *Philebus*, 55–58, 61 e, distinguishes between (exact) mathematical sciences and sciences which use little or no mathematics.

[81] Fr. 18, Diels.

[82] Thales (?), Diogenes Laërtius, I. 35; Aeschylus, *Prometheus*, 981;
Euripides, *Hippolytus*, 252; contrast Paron, quoted by Aristotle, *Physic.*,
IV. 13, 222 b. 17 f.

[83] II. 49, tr. Godley; cf. II. 81.

"It was Melampus who taught the Greeks the name of Dionysus, and the way of sacrificing to him, and the phallic procession; I would not in strictness say that he showed them completely the whole matter, for the later teachers added somewhat to his showing." Whatever one may think of his conclusion, one must admire his evident research and discrimination; however embryonic, historical science is there; for there is the observation of resemblance and a theory to explain it as well as the points of difference, and an attempt to establish the connection between cause and effect. By medical writers we are told that it was sad experience with the crude foods, which man originally used in common with other animals, that taught him the need of finding nutriment more suited to his constitution, particularly in sickness;[84] and again we are informed that the discoveries were originally made by chance, or by following the practice of animals.[85] Nevertheless one must learn from those who are able to discriminate[86] what intelligence discovers, because medicine has advanced far beyond the stage of ignorance and has made great discoveries by reasonable methods and not by chance.[87]

In the Hippocratic *Precepts*[88] we have interesting suggestions regarding the scientific procedure in medicine: "One must have regard to reasoned experience rather than to plausible argument; for reasoning is a sort of memory compounded of the reports of the senses. Sense envisages things clearly, being affected by external reality and reporting to the intelligence, which, repeatedly apprised as to the

[84] Hippocrates, *De Prisca Medicina*, 3 (I. 574 f., L.).

[85] Hippocrates, *De Affectionibus*, 45 (VI. 254 L.); *De Prisca Medicina.*, 7 (I. 584 f., L.); Galen, XIV, 674 f. Kühn.

[86] Hippocrates *De Morbis*, I. 1 (VI. 140 L.).

[87] Hippocrates, *De Prisca Medicina*, 12 (I. 596 f., L.).

[88] 1–2 (IX. 250 ff., L.).

persons affected, when, and how, retaining these reports and bestowing them with itself, remembers.   I approve of reasoning if it takes observed fact as its point of departure and methodically draws its conclusions from the phenomena; for if reasoning proceeds on the basis of what clearly comes to pass, it is found to be within the province of the understanding, which itself receives the separate data from other organs.   The organism then, we must assume, is stimulated and instructed by many and many kinds of things, by dint of force; the understanding, receiving the reports from it, subsequently develops them into truth; but if it does not start with what is certain but with specious and fictitious theory, it often leads to a grievous and painful situation . . . . I think the entire art has been perfected by observing the outcome of every condition and bringing the observations of it into a system." The text is perhaps of the late Fourth Century,[88a] but it is worthy of consideration for it expresses the ideals of the Hippocratic school.[89]

It had long been recognized that science presupposes an accumulation of observations.   Herodotus[90] assumes the existence of such records among the astrologers and seers of Egypt, "who have made themselves more omens than all other nations together; when an ominous thing happens they take note of the outcome and write it down; and if anything of a like kind happens again they think it will have a like result."   When Thucydides[91] described the

[88a] The fact that this statement agrees in some respects with the doctrines and terminology of Epicurus has led some scholars to assume dependence on him.   Of course, this does not follow.

[89] Cf. *De Prisca Medicina*, 9 (I. 588 f., L.).

[90] II. 82, tr. Godley.   We have abundant records of the kind from Mesopotamia.

[91] II. 48 ff.   His procedure and language suggest conscious dependence on medical authors or physicians whom he had consulted.

symptoms of the plague at Athens, from which he suffered, he did so in order that the scourge might be recognized in the event of its recurring; the minute records of the history of diseases given in the Hippocratic writings, which the historian evidently took for his model, were left as guides to future practitioners and as the basis for rational inference and practice. No doubt such records had long been kept in families or guilds in which an art was traditional —hence the Hippocratics insisted that an important obligation of the physician was to consult and judge the written records.[92] It must have been the same in all the arts which dealt with matters other than mere routine. On the foundations thus laid, theories were formulated. By theory one does not, of course, necessarily mean a hypothesis general and resting upon pure assumption:[93] of such there were many, no doubt, in ancient times as there are still in our day. But there were realists even in the Fifth Century B.C., whose method in practice was not very different from that of the later Empirics.

It is worth while to give a hearing to one of these, the author of the treatise *On the Old School of Medicine*.[94] "All who, on attempting to speak or to write on medicine, have assumed for themselves a postulate (hypothesis) as a basis for their discussion—heat, cold, moisture, dryness, or anything else that they may fancy—who narrow down the causal principle of diseases and death among men and make it the same in all cases, postulating one thing or two,

[92] Hippocrates, *De Diebus Criticis*, 1 (IX. 298 L.); *Epidem.*, III. 16 (III. 100 L.); Aristotle, *De Part. Animal.*, I. 1, 639 a 4 ff.

[93] For hypothesis, see in addition to the passage quoted, Hippocrates, *De Prisca Medicina*, 13 (I. 598 L.); *De Ventis*, 15 (VI. 114 L.); *De Articulis*, 10 (IV. 102 L); Plato, *Republic*, 527 a, 530 a b; Jowett-Campbell, *Plato's Republic*, II. 333 ff.; Gomperz, *Griechische Denker*, I. 237, 244.

[94] Hippocrates, *De Prisca Medicina*, 1–2 (I. 570 ff., L.), tr. Jones, adapted.

all these obviously blunder at many points even of their statements, but they are most open to censure because they blunder in what is really an art (or science), and one which all men use on the most important occasions, and give the greatest honors to the good craftsmen and practitioners in it.   Some practitioners are poor, others very excellent; this would not be the case if medicine did not exist at all (as an art or science) and had not been the subject of any research and discovery, but all would be equally unpracticed and inexpert therein, and the treatment of the sick would be in all respects haphazard.   But it is not so; just as in all other arts the workers vary much in skill and intelligence, so also is it in the case of medicine.   Wherefore I have deemed that it has no need of an empty postulate, as do matters inaccessible to observation and subject to question, about which any who would speak must use a postulate—for example, things in the sky and below the earth.   If a man were to express an opinion about these, neither to the speaker nor to his audience would it be clear whether his statements were true or not; for there is no test the application of which would give certainty.   But medicine has long had all its means to hand, and has discovered both a principle and a method, through which the discoveries made during a long period are many and excellent, and what remains will be discovered, if the inquirer be competent, conduct his researches with knowledge of the discoveries already made, and make them his starting-point.   But anyone who, casting aside and rejecting all these means, attempts to conduct research in any other way or after another fashion, deceives and is himself deceived.   His assertion is impossible; and the causes of its impossibility I will attempt to make plain by showing that medicine as an art (or science) really exists.   In this way it will be made clear that by any other means discoveries are impossible, all the more, because[95] one who discusses this art must speak of things familiar to laymen. For the subject of discussion and inquiry is none other than the sufferings of these same laymen when they are sick or in pain. To be sure, to learn by themselves how their own sufferings come about and cease, and the reasons why they get worse or better, is not an easy task for laymen; but when these things have been discovered and set forth by another, it is simple. For merely an effort of memory is required of each man when he listens to a statement of his experiences.[96] . . . . Therefore for this reason also medicine has no need of any postulate."

This pronouncement is particularly interesting because of the express recognition of the obligation to test theories by an appeal to acknowledged fact. The same author, a little farther on,[97] calls attention to the need of finding a means of determining the proper measure of food to be administered, and declares that there is no test except the feelings of the patient. His practical realism he forcefully states in opposition to the theorizers in a later chapter,[98] where he maintains that one may really learn something about nature only from the science of medicine, conceived as he conceives it. "This at least I think a physician must know, and be at great pains to know, about nature, if he is going to perform aught of his duty, what man is in relation to food and drink and his habits generally, and what will be the effects of each on each individual. It is not suf-

---

[95] I have tried here to give the sense, though it is not a true rendering of the text. The author sees in the agreement (or disagreement) of the patient with the physician the test of the latter's diagnosis.

[96] I omit a sentence. If the text is sound, the author's ideas are apparently confused.

[97] 9 (I. 588 f., L.).

[98] 20 (I. 620 ff., L.), tr. Jones.

ficient to learn that cheese is a bad food, because it gives a
pain to one who eats a surfeit of it; we must know what the
pain is, the reasons for it, and which constituent of man is
harmfully affected.   For there are many bad foods and bad
drinks, which affect a man in different ways.   I would,
therefore, have the point put thus: 'Undiluted wine, drunk
in large quantities, produces a certain effect upon a man.'
All who know would recognize this, that that is the physiological action of wine, and that wine itself is the cause of it,
and we know on what part of a man it principally acts.
Such a nicety of truth I wish to be shown in all other cases."
This is certainly a high ideal, still far from being realized; it
could hardly have been formulated, however, unless certain
physicians had at least made very close study of individual
foods and individual constitutions, checking their conclusions by further observation and questioning of their patients, in their actual practice disregarding abstruse theories,whatever they might conceive to be the relation of the
nature of man to nature in general.   One can hardly conceive of an ancient Greek of education who did not have
some theory, however much he may have kept it in abey-
ance in pursuing his practice; for Aristotle says[99] that no
one had ever proposed different principles for things eternal and things perishable, but all say that the same principles apply everywhere.   There were, as has already
been said, enough who concerned themselves with the
larger relations, as the *Epidemics* and the treatise *On
Climate, Water and Environment* clearly show.   Here
also one can not mistake the true scientific spirit, even
when data were insufficient and conclusions premature.
Plutarch[100] quotes the poet Epicharmus as saying, "Apply
the stone to the yardstick, not the yardstick to the stone,"

[99] *Metaph.*, II. 4, 1000 b 32 f.
[100] *Moralia*, p. 75 f.

and interprets this as a warning not to force nature to conform to a theory, but to make the theory fit the facts.   It
is the common Greek notion of "saving the phenomena,"
whatever theory may be propounded.   The ideal set forth
by the medical writer of the Fifth Century is that of the
later men of science, as one might show by many examples.
It may suffice to refer to Aristoxenus, who had been a Py-
thagorean and had joined the school of Aristotle.   With
such antecedents, one might have expected him to display a
fondness for abstruse, mathematical theories and formulas;
but in his treatment of music he was a rigorous scientific
empiric, ruling out all mathematical and physical theories
about the nature of sound and appealing only to the ear.[101]
Aristotle himself is commonly misjudged, because one part
of his writings—the speculative part—has been unduly
stressed.

That the associates of Aristotle in the Lyceum devoted
themselves with such energy to the study of nature in various branches of science was of course due in no small measure to his precept and example.   One might cite many
passages in illustration of his method in dealing with concrete problems; but one will suffice.   Speaking of the apparent motions of the heavenly bodies he says,[102] "That the
movements are more numerous than the bodies that are
moved, is evident to those who have given even moderate
attention to the matter; for each of the planets has more
than one movement.   But as to the actual number of
these movements, we now—to give some notion of the
subject—quote what some of the mathematicians say, that
our thought may have some definite number to grasp; but,
far the rest, we must partly investigate for ourselves, partly
learn from other investigators, and if those who study this

[101] *Harmon. Elementa*, cc. 32, 33, 44, ed. Meibohm.
[102] *Metaph.*, XI. 8, 1073 b 8 ff., tr. Ross.

subject form an opinion contrary to what we have now stated, we must esteem both parties, indeed, but follow the more accurate."   Wherever we turn in Greek literature we find the same devotion to research and the same open-mindedness to such truth as could be discovered, along with the frank confession of error[103] and the recognition of the limitations of art or science.[104]   The beginning of the *History of the Peloponnesian War* by Thucydides is a striking example of the temper of a man of intelligence in Periclean Athens, not only in the calm assessment of the psychic and physical forces involved in the struggle and in the conspicuously impartial record of events, but in the psychological insight displayed in the speeches, where with the genius of a dramatist he enters sympathetically into the situation of each party and interprets the situation from every point of view.   His reconstruction of the early social conditions in Greece from the state of backward communities is a fine bit of archaeology and bespeaks a schooled historical imagination.[105]   Well might he aspire to make a contribution to endure for ages.   Such work does not flourish in a vacuum—it can be cultivated only in an atmosphere charged with the spirit of science.   The impulse then given never spent itself until Greece was physically and spiritually impoverished and exhausted.   In the Third Century one of the great experimental physiologists expressed the enthusiasm of the inquirer into the mysteries of nature:[106] "Those who are altogether unaccustomed to research are at the first exercise of their intelligence befogged and blinded, and quickly desist for fatigue and lack of intellectual power, like those who without practice at-

[103] For example, Hippocrates, *Epidem.*, V. 27 (V. 226 L.).
[104] Hippocrates, *De Arte*, 8 (VI. 12 f., L.).
[105] Cf. G. F. Abbott, *Thucydides: A Study in Historical Reality*, Ch. III.
[106] Erasistratus, quoted by Galen, *Scripta Minora*, II. 17, ed. Müller.

tempt a race; but one who is accustomed to investigation, worming his way through and turning in all directions, does not give up the search, I will not say day or night, but without resting his whole life long. He will turn his attention to other and still other things not alien to the subject under investigation until he arrives at the solution of his problem." It was such high devotion and noble enthusiasm as this that Euripides[107] celebrated: "Blessed is he who has taken knowledge of science, having no impulse to his fellow's harm or unjust deeds, but contemplating the ageless order of deathless Nature—how it came to be formed, its manner, its way; over such comes no care for deeds of shame."

Plato[108] several times makes Socrates express similar views regarding the devotees of philosophy, and indeed the picture of the ideal philosopher and scientist which was conceived in the Fifth Century continued to hold the imagination of the Greeks to the end. One recognizes the leading features in the *Prometheus* of AEschylus, the Solon of Herodotus, the Socrates of Plato, and the Democritus and Archimedes of later story; everywhere it is the unselfish devotee of Truth, who sacrifices his own pleasures and material interests in his absorption about his real business, whether he chance to regard it, like Socrates, as God's business, or, like the physicians, as the business of humanity. The Hippocratic writings hint that one's love of his science necessarily involves a love of mankind;[109] and Thucydides[110] records the great mortality among physicians who attended, though to no purpose, the victims of the plague at Athens. One thinks of the devoted priest or pastor, as I recall with pride the reckless disregard of his own life which my father

---

[107] Fr. 910, Nauck.
[108] *Theaetetus*, 172 d–176 a, *Republic*, 500 c–d.
[109] Hippocrates, *Praecepta*, 6 (IX. 258 L.).
[110] II. 47.

displayed in my childhood during a visitation of the plague. A Hippocratic writer goes so far as to say that when there are several ways of achieving the health of the patient the least troublesome is both the most scientific and the most worthy of a good man.[111]   The physician, like all scientists, will pay less regard to wealth than to fame and virtue.[112] The Hippocratic *Oath* and *Law* formulate an ideal which in essentials holds even to-day.

Although Aristotle rarely rises to real eloquence in his extant works, he was much admired in antiquity for his style in less technical works, which have for the most part been lost or are represented at best by scant excerpts.   A few passages in his extant writings give the impression of having been taken over from more popular treatises written in the enthusiasm of his youth.   Of that number perhaps is the following[113] which deserves to be quoted, although it is familiar to all lovers of Greek literature.   "Of things constituted by nature," he says, "some are ungenerated, imperishable, and eternal, while others are subject to generation and decay.   The former are excellent beyond compare and divine, but less accessible to knowledge.   The evidence that might throw light on them, and on the problems we long to solve respecting them, is furnished but scantily by sensation; whereas respecting perishable plants and animals we have abundant information, living as we do in their midst, and ample data may be collected concerning all their various kinds, if only we are willing to take sufficient pains.   Both departments, however, have their special charm.   The scanty conceptions to which we can attain of

[111] *De Articulis*, 78 (IV. 312 L.).

[112] Hippocrates, *Praecepta*, 4 (IX. 254 f., L.); Galen, I. 60 f., ed. Kühn; cf. Aristotle, *Polit.*, I. 11, 1259 a 6 ff. and Apuleius, *Flor.*, 18, for the stories about Thales, the sage *par excellence*.

[113] *De Part. Animal.*, I. 4, 644 b 22 ff., tr. Ogle.

celestial things give us, from their excellence, more pleasure than all our knowledge of the world in which we live; just as a half-glimpse of persons that we love is more delightful than a leisurely view of other things, whatever their number and dimensions. On the other hand, in certitude and in completeness our knowledge of terrestrial things has the advantage. Moreover, their greater nearness and affinity to us balance somewhat the loftier interest of the heavenly things that are the objects of the higher philosophy. Having already treated of the celestial world, as far as our conjectures could reach, we proceed to treat of animals, without omitting, to the best of our ability, any member of the kingdom, however ignoble. For if some have no graces to charm the sense, yet even these, by disclosing to intellectual perception the artistic spirit that designed them, give immense pleasure to all who can trace links of causation and are inclined to philosophy. Indeed, it would be strange if mimic representations of them were attractive, because they disclose the mimetic skill of the painter or sculptor, and the original realities themselves were not more interesting, to all at any rate who have eyes to discern the reasons that determined their formation. We therefore must not recoil with childish aversion from the examination of the humbler animals. Every realm of nature is marvelous; and as Heraclitus, when the strangers who came to visit him found him warming himself at the furnace in the kitchen and hesitated to go in, is reported to have bidden them not to be afraid to enter, as even in that kitchen divinities were present, so we should venture on the study of every kind of animal without distaste, for each and all will reveal to us something natural and something beautiful." It needed not this word to teach us that Aristotle worked *con amore* in all that he did; but it is heartening to find him so truly Greek in his sustained enthusiasm and in his will to find beauty, as well as truth, in everything.

The Hippocratic treatise *On Decorum* draws an ideal portrait of the man of science—especially the physician—and the philosopher and contrasts them with the charlatan, who is depicted in the colors familiar to all in the Platonic picture of the Sophists. Whereas elsewhere[114] physician and philosopher are distinguished in respect to method, they are here said to have the same virtues and to differ but little. The points of resemblance are chiefly in regard to character, but it is obvious that the author had also in mind their similarity in outlook and attitude toward their functions. He goes so far as to declare that the physician is a god-like philosopher and that medicine must be carried into philosophy and philosophy into medicine.[115] Though this treatise is very probably late, the point of view is not. Aristotle says:[116] "It behooves the natural philosopher also to examine the first principles of health and disease, since neither health nor disease can come to things deprived of life. Therefore, one may fairly say, the majority of those who pursue the study of nature come finally to questions of medical science, and those physicians who take a more philosophical view of their science have to deal with philosphy." Again he says,[117] "Health and disease are subjects for the consideration not only of the physician, but also, up to a certain point, of the natural philosopher. We must not overlook their differences and the different manner in which they regard their different provinces, though what commonly happens bears witness to the fact that their fields are adjoining; for physicians of culture who are of an inquiring mind discourse to a certain extent about Nature (natural science) and claim to derive their principles

[114] Hippocrates, *De Prisca Medicina*, 1 (I. 570 L.), *De Natura Hominis*, 1 (VI. 32 ff. L.).
[115] Hippocrates, *De Habitu Decenti*, 5 (IX. 232 L.).
[116] *De Sensu*, 1, 436 a 17 ff.
[117] *De Respiratione*, 21, 480 b 22 ff.

from it, while the cleverest natural philosophers generally lead up to medical principles."

Just why the ideal physician should have been pronounced a god-like philosopher is not quite clear. It may be that the author merely had in mind "divine philosophy," as we have been taught by Plato and Aristotle to regard it; but he may also have had more special reasons for bestowing the honorific title, which seems to us so extravagant. He had just been emphasizing the fact that the true physician combines theory and practice; and since he was comparing and contrasting the physician with the natural philosopher, it is possible that he was thinking more especially of the difference between them—the *physicus*, dealing chiefly with matters whose causes and laws he might investigate, but being human could not control; whereas the physician, studying the causes and laws of health, was, like a god, able to effect the thing with which he was concerned. This is at least an intelligible thought, however remote from the humble attitude which Xenophon attributes to Socrates.[118]   We must recall the high hopes which physicians then entertained of the advancement of knowledge.   Medicine, as we have seen, was thought to have attained an established principle and method in accordance with which one who was properly equipped might achieve all that remained to be discovered.[119]   One writer even thinks that perfection has already been reached, since there is nothing left to be found out.[120]   If Socrates[121] rated for fools those who might dream of employing the laws of nature which they studied in the production of winds or rains, Empedocles seems to have entertained even more extravagant

[118] Xenophon, *Memorabilia*, I. 1, 15.
[119] Hippocrates, *De Arte*, 1 (VI. 2 L.).
[120] Hippocrates, *De Locis in Homine*, 46 (VI. 342 L.).
[121] Xenophon, *Memorabilia*, I. 1, 11–15.

hopes.[122] One who knows how certain discoveries, which to us seem quite commonplace, have fired the imagination of scientists in all the ages since and have led them to expect the ultimate consummation in the near future, may smile ruefully at the indefeasible optimism of man, but, remembering Aristotle, Bacon, and Descartes, he will be slow to cast a stone at these ancient enthusiasts. There is, moreover, another line of thought which may help to explain the epithet "god-like" given to the physician. The science of medicine, then as now, included not only diagnosis but prognosis. To predict the future has always been regarded as the supreme test of science. The Greeks were especially devoted to the practice of foretelling the course and issue of a disease[122a] and the more sober-minded endeavored, as in the Hippocratic *Prognostic*, to divest it of all that was theatrical, while recommending it as the best means of winning the confidence of the public. From the earliest times, apparently among all peoples, prophecy was regarded as the function of the sage, and typical anecdotes regarding early Greeks, such as Thales, Anaximander, and Pherecydes, show how wide-spread was the belief that the wise man (or "divine man") had this gift. Pliny[123] attributes it to their *divinitas*; and Aulus Gellius[124] quotes Favorinus as saying that the chief difference between gods and men would disappear if men had prescience of the future. In the same vein Diogenes Laërtius,[125] who attributes a number of prophecies to Epimenides, says, "Some declare that the Cretans offer sacrifices to him as a god; for they claim that he had the gift of prophecy."

[122] Fr. 111, Diels.
[122a] Hippocrates, *Prognosticum*, 1 (II. 110 L.) advises the physician to practise prognosis.
[123] *Nat. Hist.*, II. 191.
[124] XIV. 1, 6.
[125] I. 114–115.

For our purpose it matters little just what notion, or complex of ideas, suggested this description of the ideal man of science.   What is quite clear is that men of intelligence had come to regard the knowledge of Nature and its ways as far advanced and on the road to speedy perfection.   In the retrospect the claims and hopes of that time must appear childish and grotesque, as must those of any time to men who come after and realize that Nature is vastly more complex than had been supposed; but to dismiss such thoughts as mere occasions for condescending good humor were to miss entirely their real significance.   Such claims clearly indicate that the questions which had been put to Nature had been satisfactorily answered, or appeared to be just awaiting the true answer.   The intellect had raised its problem, and was finding the solution.   At bottom the operations of the human mind have not changed; what has changed with the process of the ages is the definition of the facts to be explained and the specific means of clarifying and testing the ideas which suggest themselves as accounting for them.   To trace that development is to reconstruct the history of thought, of which the history of science is but a part.

PART II. METHODS

I. INTRODUCTORY

Greek science has not, as a rule, been fortunate in either its assailants or its champions.   Too often they have shown themselves wanting in the historical sense which is indispensable in setting a true value upon anything that belongs to the past.   There are those who were severely critical of its shortcomings, while others have claimed too much for it, and have in consequence played into the hands of their opponents.   Perhaps the champions, besides winning our admiration for their generous impulses, were more nearly in the right, though the historian, bent on sympathetic understanding, will be chary alike of praise and blame; the fault in either case is apt to lie in the application of a false standard.   Ancient science, like ancient philosophy, must not be too closely compared with the modern; if comparison there is to be, it should in justice be made with contemporary or antecedent thought.   This, however, is neither possible in detail, nor would it be very profitable because of the almost total loss of contemporary documents of other peoples and of the extraordinary difference in this regard between the Greeks and other nations of antiquity. It may be said, and truly said, that Greece by its singular eminence in this and other respects directly challenges comparison with the achievements of modern times; and perhaps it is both inevitable and right to bring out the points of similarity and difference, provided one is at the pains first of all to understand the phenomena in the light of history.

Understanding in such a matter is conditioned by many things, particularly by a due regard for the operations of the human mind and by the cultivation of the imagination necessary to reconstruct a whole from fragmentary evidence. The latter need is in some directions so apparent that one might think it superfluous to dwell upon it; but an acquaintance with certain well-known works, whose influence may be traced almost everywhere in the literature relating to the history of science, will convince the student that its importance is in fact little recognized. History, like every branch of science, requires imagination both to suggest possible connections and to draw the probable or necessary inferences from the available data. The too literal-minded will see at best the recorded facts, where the man who inquires into their meaning will discover important implications. This of course involves reading between the lines, which calls for caution; but it is as necessary as it is fruitful. Let us suppose that an archaeologist discovers balances with both equal and unequal arms—it is then evident that the principles of the lever were at least empirically known to those who constructed the balances. If these are marked with a graduated scale, and if the weights are standardized, one must conclude that the theory also was known. The disposition so characteristic of our time to restrict the use of the word science to the "exact," or mathematical, sciences has along with its advantages the unfortunate consequence that many persons come to regard the mental operations as essentially different in the several fields where they find employment, whereas they differ only in respect to their subject-matter or to the degree in which it has thus far been found possible to define the data with which they have to deal. Science is still, as it always has been, in intention one—in certain directions it has been immensely advanced by ignoring every

aspect of nature except that which lends itself to mathematical or mechanical statement. Perhaps in time things
that now seem intractable will yield to this procedure; but
whether they do or not will make no difference either to
their importance or to the intellectual operations that deal
with them.

It can not be too often or too strongly emphasized that
our whole spontaneous way of regarding nature has come
down to us from pre-historic times, and that the procedure
of science is nothing but a refinement of the processes which
come into play in ordinary life. One may cite legal practice as an illustration. No doubt the earliest decisions were
reached, to all appearances, by the native mother-wit of the
judge; improvements came by the method of trial and
error, sad experience suggesting needful precautions and
more extended investigation. In the end there was formulated the law of evidence. Here, as elsewhere, it is true
that "the gods did not reveal all things to mortals from the
beginning, but by searching in time they find the better."
In the recognition of this fact is the beginning of historical
wisdom. In fact, it is quite impossible to understand
ancient science, not to speak of doing it justice, without
taking this point of view.

In the preceding chapter attention was drawn to certain
aspects of Greek scientific and philosophic thought which
are seen in a clearer light when they are regarded as points
of view inherited from a pre-scientific age. As we proceed
we shall have occasion to refer to more instances of the same
kind. At this point it seems desirable to advert to a number of facts, well known to psychologists, which may help
us to approach the study of Greek methods in science in a
more open-minded way and make it easier to understand
what one finds in the records, and why one does not find
some things that we miss.

In the first place, then, we must observe that our mental processes were originally, and are still for the most part, dominated by practical rather than purely theoretical interests. They are rough tools rather than instruments of precision. Of course they are capable of great technical improvement, and when the need arises, and attention is concentrated on the process rather than the general results, man is not slow to devise means of satisfying his requirements. The dead weight of inertia is such, however, that the need must be very pressing before anyone bestirs himself to meet it; once the search has been started, it may become a sport followed for its own sake, as every interest of man, when it has grown sufficiently to stand alone, tends to become a discipline or an institution. Thus deductive logic was the inevitable answer to the challenge of the Eleatic dialectic and the quibbles of the Eristics—beginning, one may say, with Socrates, in essentials completed by Plato, it was systematized by Aristotle as the perfect instrument of the elenchus. But its very perfection as a system for the ready discovery of sophisms, as a means of criticism of propositions in their formal aspects, while it reveals the interest to which it owed its origin, tended to obscure its incompleteness as an instrument of thought; emphasizing the testing of propositions, it naturally took little account of the process by which propositions are framed—in other words, of induction, the method of inferring a truth from given data. To the natural man this phase of the intellectual process presents no difficulty, because he is constantly drawing conclusions from facts which seem to him obvious. Inductive logic is perhaps inevitably a later product of the scientific interest.

This becomes even clearer when one regards the behavior of the mind in purely practical affairs. There is really no difference in principle between framing a theory and setting up an end or ideal. In either case one starts with the given and works toward an objective, at first vaguely visualized. It is the end that dominates the process throughout. Just because the given is taken for granted it will not be specially scrutinized unless the means prove utterly inadequate to the end to be realized. In that event the means at one's disposal have to be reexamined and perhaps reconstituted, or the proximate end must be suited to what one now sees the given is good for. So, in an affair which one has very much at heart, there is constant interaction between means and end, one being adjusted to the other; but, if the person be an idle day-dreamer, content with merely envisioning an ideal without seriously striving to attain it, the terms may continue to stand apart instead of functioning together. Now it is obvious that in the beginning science was a pastime, a sort of intellectual game which had a great fascination for men like the Greeks, possessed of a lively curiosity. Theories naturally sprung up among them in profusion as, according to their thinkers, animal forms were gendered in the heyday of the world's youth. Small wonder that some of them were monsters. The difficulty lay not in the reasoning, however, which was essentially the same as any one now can perform, but rather in the premises from which they started. In particular it was the body of data they had at their disposal. Just because they were given, because they were assumed to be just what they gave themselves out to be, they were taken for granted with only a cursory survey. What Mill says[1] is apparently quite true: "Every branch of natural philosophy was originally experimental; each generalization rested on a special induction and was derived from its own distinct set of observations and experiments." We shall see that the Greeks were not wanting in observation nor even in experiment, and that the theories or generalizations which

[1] *Logic*, Bk. II, Ch. IV, §5; Bk. III, Ch. IV, §2.

they framed were the reasonable inferences from the data they worked with; and we shall be able to point to a very large number of experiments which are either expressly attested or with certainty to be inferred from assured facts. When one contrasts Greek and modern science in regard to method the difference lies, almost entirely, in the greater care which the modern scientist is expected to exercise in determining the precise nature of his data. It is at this point that the "experimental method," of which one hears so much, has demonstrated its supreme value. Perhaps one may say that fundamentally this difference between ancient and modern science is due to the greater seriousness with which it is now pursued; science has become a business —in some ways, the chief business—of the world today, instead of being, as Aristotle thought, a luxury of a people possessed of wealth and leisure. When science is the basis of an elaborate industrial system, which supports the entire frame of civilized society, it is a matter of consequence whether or not the facts are what they are supposed to be, and whether or not facts and theories function together.

In view of the great natural interest in explanation—in other words, in generalizations and theories—it may at first sight appear strange that, whereas Aristotle virtually brought to perfection deductive logic, the logic of induction, which aims to reduce to a system of rules the process of inferring generalizations from particulars, was far less advanced in ancient times. The most frequent criticism of the Greeks in matters of science is precisely this, that they were too ready to advance theories; and the reason for their failure is commonly alleged to be that they reasoned from too scant an array of facts. This may be true in certain fields, but it is not true of all; for in regard to matters falling within the provinces of psychology, ethics, and politics their data were ample and their generalizations such that the best modern thought has not got much beyond them. In certain physical sciences it is true that their theories were very unsatisfactory; but the explanation of their shortcomings is equally inadequate. Owing to the fact that many fields of nature were still unexplored, there was of course no such wealth of data at hand for the investigator as are now available; but valid conclusions may be drawn from something less than a complete survey. Frequently false inferences were deduced from observed fact, simply because of analogies which have since been found to be misleading, as when the bursting of a bottle, when water in it freezes, was explained by the hypothesis that water contracts when it passes from the liquid to the solid state, leaving a vacuum, so that the walls of the bottle are not able to support the atmospheric pressure.[2] This inference was in fact a brilliant generalization, suggested no doubt by many observations, such as the contraction of molten metals poured into molds. Chemists tell us that water is the sole exception to the rule, and that at the critical temperature there seems actually to be a momentary contraction preceding the expansion. In such a case the cause of failure to draw the right conclusion must be found elsewhere. The reason why the logic of induction was so slow in developing is rather that the process seemed to be self-evident, because it was performed so easily and naturally. The method of science is well described as the inductive-deductive, which is only another way of describing the functioning together of means and ends in practical affairs. An attempted deduction from a false hypothesis will inevitably bring out negative instances, as the effort to realize an ideal will reveal the inadequacy of one's resources, if they have not been truly assessed.

[2] Cf. E. Wiedemann, "Ueber das Experiment im Altertum und Mittelalter" (*Unterrichtsblätter für Mathematik und Naturwissenschaften*, XII (1906), 122 f.).

The characteristic which is thought to distinguish Greek from modern science is its deductive procedure. If one takes Aristotle as the typical representatives of Greek method there is much to be said for such a view; for he lays it down that it is the business of the particular sciences to develop by deduction from the highest principles the specifications, differences, and functions of their several kinds, so far as they are permanent and essential.[3]   To be sure, this is the ideal also of modern science; there is, therefore, no difference in principle, but only a change of emphasis, since Aristotle, no less than his modern critics, recognized that the highest principles, which do not admit of being proved by deductive methods, must be won by induction. So far as criticism of his procedure is justified, it relates to his method of induction; and here, in fact, there is ample ground for dissatisfaction, because the process as it is set forth in the logical treatises is indeed rather perfunctory.[4] Psychologically this fact may be explained, as has already been suggested, by the ease with which one naturally generalizes, whereas the need of caution in deducing consequences from general principles is quickly brought home to one.   Historically it is no less clear why the treatment of the two processes of reasoning received so unequal attention; for logic grew out of disputation and dialectic, which dealt, not with concrete facts, but with general ideas. Moreover, Aristotle took over the problem from Socrates and Plato, who had developed induction to a point which must have seemed all but final; though this occurred chiefly in the field of ethics, where the primary data are not mere brute facts but ideas, themselves the product of induction performed for the most part in the immemorial past.   As such they did not specially invite further scrutiny; they

[3] See the passages cited by Maier, *Syllogistik des Aristoteles*, II. i, 399 f.
[4] *Ibid.*, 370 ff.

were given, and must be accepted. What remained for the reasoner was to organize them, as found, into unobjectionable principles, which could be applied in particular situations. Neither Plato nor Aristotle ever lost sight of the need of making the general ideas, or ideals, function in the ordering of concrete reality, of controlling the actual by the principles inferred; but in natural science and in ethics alike they were hampered by the too ready acceptance of the terms within which reasoning operates—the primary data and the highest generalizations—as given, and therefore not to be further questioned. Where they fell short was not so much in theory as in practice, a failing only too common at all times.

What is true of Aristotle's logic is not necessarily true of his procedure in every field of inquiry. There are subjects which he treated in the manner one might expect from the general complexion of his theory, as sketched above; but there are also extensive fields, in which one is amazed by the minuteness and variety of his observation of fact. The former have to do for the most part with matters in which, as the Hippocratic writer said, hypotheses were necessary, because there is no possibility of appealing to indisputable fact; whereas the latter fall to the empirical sciences. Herein also Aristotle was the heir of the past. Aside from dialectic, which was formulated under the influence of Socrates and Plato, there were two principal scientific interests which had been developed to a stage where they must exercise a marked influence on any thinker of the first rank. One of these was mathematics, which is by nature abstract, and, as it was then cultivated, called for detailed observation nowhere except in astronomy. Plato coupled mathematics with dialectic, making the former a propaedeutic to the latter, and even proposed that one forego observation in astronomy and treat it in a purely abstract and deductive fashion. Just how this suggestion was intended is doubtful, because there is good reason to believe that he set problems in astronomy, the solution of which necessarily called for observation; but in any case the nature of mathematics and the relation in which it was brought to logic must have exercised a powerful influence on his pupil, who remained to the end essentially a Platonist. The other interest centered in medicine, but had long embraced not only general biology but meteorology also. As the son of a physician, Aristotle may be assumed to have shared this interest from childhood, and everything goes to show that it deeply influenced his thinking throughout his life. Medicine was then, however, as it has remained, an art rather than a science, and its aim was chiefly practical. Success in the practice of this art has always depended more on close observation of details than on theory, and it is doubtless due to the established medical tradition that Aristotle displays in his biological treatises not only the astonishing capacity for minute observation, but also the loving devotion to it in practice which have commanded the admiration of all competent students. The loss of the great body of medical literature contemporary with Plato and Aristotle, which we know once existed, renders it impossible to estimate the latter's debt to it, but it may be taken for certain that it was very great.

What strikes one at once is the contrast presented by Aristotle's procedure in these two fields. He could not bridge the gulf which separated them any more than he could in logical theory bring deduction and induction into such relation as to make them function effectively together. The contrast is in fact almost as great today as it was in his time; for biology has not yet made good its claim to be reckoned among the exact sciences. As has already been said, mathematics had already attained a high development; but it was abstract, and it was as geometry chiefly that it was cultivated. As such it found application in relatively few relations, and there only to a limited extent— at first in astronomy, later in mechanics. Mathematics has slowly extended its domain, as an ambitious land-owner acquires contiguous fields; between the provinces of geometry and biology the distance was too great to be cleared at once. Intermediate sciences, such as chemistry, alone seem capable in time of bringing them together. As it was, mathematics in the later centuries of Greek thought won great triumphs in astronomy, mechanics, and optics by the union of theory with observation. When at the Renaissance science was reborn, it laid hold of these very acquisitions of the ancients, somewhat enlarged by Islamic thinkers, and with them assailed the current Aristotelianism, which was based almost entirely on the works characterized by the deductive method. It is interesting to note the prominence of men trained in medicine in this assault on deduction in the name of induction; one Aristotle was arrayed against the other.

A great part of the criticism of the ancients argues chiefly the want of schooled imagination on the part of the critic, who has often been at no pains to visualize the conditions under which science grows. A scholar, who is a notable exception, has said,[5] "In every science the steps by which one rises successively to the highest conceptions, to the most brilliant discoveries, depend on certain conditions which one can not always calculate in advance, but which one recognizes by a careful study of history; almost always the ascent is long and difficult . . . . No human genius can, in a science that requires observation and experiment, pass the limits imposed on him by the available instruments." This important observation is as true of intellectual as of

[5] Daremberg, *Histoire des Sciences Médicales*, I. 18.

physical instruments.   In the preceding paragraph attention has been drawn to the handicap of Aristotle, owing to the logical method which he inherited.   As regards physical apparatus, he was probably in even worse case.   One who is somewhat acquainted with the remarkable achievements of the Greeks in art and literature, which are on a level with the best that man has ever attained, would be apt to imagine a background in other respects comparable with that of our time.   Even the careful historical scholar finds it difficult to imagine the conditions under which one worked in the Fifth and Fourth Centuries B.C.   It is certain, however, that the means employed by the great artists were extremely simple, though skilled hands found them adequate for their purposes.   The fact is that Greek civilization was as little mechanical as was possible, considering the results.   Instruments of precision were generally wanting, and so long as one could dispense with them, they were not required.   Furthermore, life in general was far less complex that at present, and consequently there was no such specialization in industry as now exists.   One may think that this fact does not much signify, because there was evidently enough skill available to produce any instrument that a scientist might require.   No doubt that is true, but one must first feel the need of a thing and clearly conceive the way in which it is to be employed, and that is not so simple as might be thought.   Of course, the genius does just that; but mechanical genius is rare at best, and especially rare in an environment that does not habitually think in mechanical terms.   Investigation would probably show that by far the largest number of instruments which have been employed in our time with such success in scientific experimentation were taken over either bodily, or with slight alterations for greater precision, from the workshops of industry.

Another fact to be considered in the progress of science is the number of workers and the possibility of ready communication between them. This is a matter to which little attention is paid, although it is of the greatest importance. Try to realize the contrast between conditions in ancient Greece and in the modern world, so far as it participates in the advancement of science, and one may appreciate what was accomplished by the pioneers. At present the workers in pure science and in science in the service of industry number hundreds of thousands, and they are in close communication, not only by personal association in laboratories and frequent larger meetings but more especially by the publication and prompt tabulation of methods and results, whereby it becomes possible for almost anyone sufficiently interested to keep abreast of current researches. In ancient Greece it was not so. We are apt to think of Greece as a whole and, reckoning the total population, to think of the proportion of those who contributed to knowledge as perhaps not very different from that obtaining today. But without insisting on the difference between initiating a thing and carrying it on after it has become established, one can readily show that the comparison would not be fair. It is certain that in all the higher phases of civilization the number of communities in Greece that had a share in its achievements was very small—a dozen towns of very moderate size would have to be included, whereas the remainder of the population was as little concerned as Albania, Turkestan, and Tibet today. The occasional man of science, who in some miraculous way was fired with enthusiasm for research, nursed the flame in isolation or at best in company with a small coterie of like-minded men, comparable to Franklin's Junto or the Gresham College Society. As for communication between individuals and groups, it was, before the end of the Fifth Century, largely a matter of chance whether one thinker became acquainted with the published work of another. That thinkers did, nevertheless, somehow learn what others were thinking and saying, as one must infer from the record, bears witness to an interest which does great credit to them. In the schools founded by Plato and Aristotle, and later in the institutions at Alexandria fostered by the Ptolemies, larger numbers met and cooperated, for the most part freely; for they were not in general regimented as they now are in some places, where research is too much organized, but were left free, as the creative mind must be left, to play with the matters they happened to take in hand, with such help and suggestions as they might get from their fellows. The same plan is happily now being adopted in some of the great industrial laboratories for research, but with far greater numbers and with facilities limited only by the requirements of the workers. One important consequence of this closer cooperation and communication is the prompt checking and extention of the methods and results of investigation. One who works in isolation is always likely to be led into error by chance conditions, which may be detected only by the repetition of the experiment (with variations) by other men working in other surroundings; moreover, the investigator is apt to see only what he is looking for, and that will depend on his present interest, while those who repeat the operation will probably have other interests and will consequently discover further conclusions to be drawn from the facts observed. Thus laws of limited scope are confirmed and, through their application to wider fields, generalized with the necessary restrictions of statement. Discoveries are generally made by individuals either by chance or by the intuition of genius; they are commonly developed and applied by lesser men in greater numbers. In Greece, with its limited number of participants, working in complete or comparative isolation, a surprisingly large contribution was made in the discovery of facts and methods, but there was wanting the army of workers necessary for the checking, extension, and application of those discoveries.

As has already been said, discoveries in science are partly due to the inspiration or intuition of genius, partly to chance, if one may properly speak of chance in a world where purpose rules. What one means, of course, is that the thing found was not expected and could not be anticipated by the finder. After a certain group of phenomena has been long studied and reduced to order it becomes possible to anticipate the results of further investigation, because the known facts create a presumption that elements in certain regards related will show similar reactions. This stage has been reached, for example, in certain fields of chemistry; but even in fields far less explored the investigator, like the clever detective, may on slight or even on no clear evidence conjecture what search will bring to light. Any scholar must be aware of this faculty of the mind, which suddenly and without warning confronts him with a theory, and then rationalizes its action by citing facts which support it, but which may never have been consciously brought together and seen in relation to one another. The man of scientific habit and training will of course hold the hypothesis provisionally and accept it only after it has been tested; while others may regard it as final. It is much easier, however, to criticize an idea than to originate it; and, in the last analysis, the presentation of a new generalization, even if it has to be revised, and the construction of a new method, though the proponent may little realize its possibilities, are more important than the determination of individual facts.

The Greeks have been severely criticized for the form in which they stated their views on scientific subjects, or perhaps one should rather say that the manner of presentation is responsible for much unfavorable comment on their methods. One point often made against them is that they proceeded deductively in their exposition, beginning with general propositions and inferring from them the explanation of particular facts. This criticism does little credit to those who offer it; for it is hardly more true of ancient than of modern scientific treatises. The ideal of science is now, as it always has been, to derive the particular from the most general principles; and it is still the common practice of men in all departments of knowledge to present their conclusions in that manner. There are various reasons for this procedure. Aristotle often distinguishes between things more knowable in themselves, meaning universal propositions, and things more clear to us, meaning concrete objects; and, as we have already hinted, he was probably influenced by the method followed in mathematics. Another consideration deserves mention. Wisdom, to which philosophy and science felt themselves akin, has always tended to express itself categorically, rather in the manner of the prophet who speaks with authority. It may be the formulation of the experience of a long life, but no reason is given, or if one be given, it is as likely as not to take the form of a trivial example. Moreover, Greek science was inherently aristocratic; and there is something in the observation of William James[6] that "the gentleman gives nothing but results, where the vulgarian is profuse of reasons." It goes without saying that the deductive and summary statement does not necessarily imply the absence of a foundation in good reasons and abundant observation, though a thoughtless reader may so conclude. Mill was clearly right in saying that every branch of science was originally experimental; it is for the historical student to discover,

[6] *The Principles of Psychology*, II. 371.

so far as he may, how accurate and extensive the experimentation was in any given case.

Just there lies the greatest difficulty for the historian and perhaps the greatest danger for science. Who but one, if such there be, who knows the entire history of chemistry could remotely guess what infinite labor is summed up in the simple-seeming statements of even an elementary text? From the fact that generalizations are false or inadequate, it is not safe to conclude that the reasoning which led to it was *a priori*. One writes in general to convince the intellect, and the deductive is still regarded as the only method that is truly apodictic. In the case of a modern treatise one does not require that the author shall reveal the process by which he arrived at his conclusions—he and his readers are generally best content when his new-found truth is shown to be the logical consequence of long familiar truths. The form of exposition is in most cases the reverse of the order of discovery. Of course there are short cuts, as when one reasons by analogy, starting with a few observed similarities and arriving at a theory, clearing the interval between fact and law at a leap. This, which probably is the most common process, is an abbreviated induction; the interval which was so summarily passed has then to be bridged by deduction. This is the so-called analytical method. It is perhaps just because this method has proved so fruitful in every branch of science that the deductive form has so captured the imagination of men. It is to be doubted, however, whether it is really the best for the advancement of knowledge, which must ultimately depend upon the appreciation on the part of men generally of the truth and the method of science. The deductive form of presentation may mean something to the man who has gone through the necessary preliminaries and may even convey to one of ordinary intelligence the conclusions of science;

but, though it is still affected in education, one finds it hard
to conceive how it can serve to train a new generation of
scientists.   If there should occur an interruption in the
line of living tradition, and if the records of actual research
should be lost, leaving only the finished treatises—which
is virtually what occurred at the close of Greek history—
what chance would there be of a ready resumption of scientific work or even of an understanding of what has now
been accomplished?

But it may be objected that all this is beside the point,
because there is no reason to believe that it applies to Greek
science.   The answer is that it is an unwarranted assumption that ancient science differed in principle at any point
from that of today.   In the succeeding chapters it will be
our task to present such definite evidence as we can of the
falsity of this prejudice; for prejudice it unquestionably is.
One does not have to go to ancient times to find parallels.
We know that Roger Bacon experimented extensively and
devoted much time and money to the construction of instruments for purposes of research.   Is there any single
instrument left to bear witness to his efforts, or any detailed
information regarding his experiments?   There is, moreover, more than a fair presumption that he was not the
only man then so engaged; but time has obscured the facts
and has left us in the dark.   Even as regards the middle of
the Seventeenth Century the situation is much the same.
At Wadham College, Oxford, a group of eminent men, including Wren, Boyle, and Wallis, busied themselves with
research and methodical experimentation.   "Unfortunately
no laboratory notes of that time are preserved, if indeed
they were made; and of many series of interesting experiments we hear but by chance allusions."[7]   We are, all of us,
so engrossed in the results that it requires a distinct effort

[7] Allbut, *Greek Medicine in Rome*, 509.

to recall how we have arrived at them. Let any scholar attempt to trace the steps by which he reached a solution of even his most important problem; unless he has kept notes in great detail of his investigations he will generally be quite at a loss, or else he will present a story which bears little resemblance to his groping course. The most conscientious scientist will generally keep a full record only of his successes, omitting his failures, which may well have taught him most. Especially among a people like the Greeks, who tended to make literature of whatever they wrote, rough notes were little likely to survive. It is only by the grace of special good fortune that any have been saved; the clinical records of the Hippocratic physicians are a notable instance, and their preservation is clearly due to the medical fraternity, which prized them as a storehouse of experience. For other examples we have to go to Archimedes and Galen.

The elaborate structure of evidence which the scholar or the scientist erects is only the scaffolding of the artisan, extremely useful in the completion of his edifice, but distracting and offensive to the artist, who prefers to contemplate the whole, defined only by the harmonious lines that give it character and meaning. A science is essentially a work of art, and the satisfaction one derives from a survey of it is essentially aesthetic. Stripped of the scaffolding, it suggests nothing of the toil that went to the making of it. The impression it leaves upon the mind is much the same as that made by the great pyramid of Egypt on the Greek historian:[8] "Most wonderful of all, though preparatory works of such magnitude were constructed in the desert, not a trace remains of the ramp nor of the working of the stones, insomuch that one might fancy that it was not builded bit by bit by the labor of men, but was set by some

[8] Diodorus Siculus, I. 63, 7.

god as a completed structure all at once in the surrounding sands."

The thought, frequently expressed by Aristotle, that while sensation presents objects for us more easily knowable, the abstractions of the reason are in themselves more knowable and more capable of being imparted, though they may be more difficult to apprehend, was not peculiar to him, but seems to have dominated science from the beginning. If the method of exposition to which it gave rise did not originate in the pursuit of mathematics, at any rate mathematicians give us perhaps the best examples of it. Used as they are to deal with abstractions they seem to feel most at home in them, and, as Plato said, to have clear vision in that high altitude while they are apt to grope helpless in the shadows of the cave, where most men dwell. "Bowditch, who translated and annotated Laplace's *Mécanique Céleste*, said that whenever his author prefaced a proposition by the words 'it is evident,' he knew that many hours of hard study lay before him."[9] The Greeks not only created the mathematical sciences, but originated and brought to perfection the method of exposition which has ever since been employed. "Thus we read in Archimedes proposition after proposition, the bearing of which is not immediately obvious, but which we find infallibly used later on; and we are led on by such easy stages that the difficulty of the original problem, as presented at the outset, is scarcely appreciated. As Plutarch says, 'It is not possible to find in geometry more difficult and troublesome questions, or more simple and lucid explanations.' But it is decidedly a rhetorical exaggeration when Plutarch goes on to say that we are deceived by the easiness of the successive steps into the belief that anyone could have discovered them for himself. On the contrary, the studied simplicity and the per-

[9] James, *The Principles of Psychology*, II. 370.

fect finish of the treatises involve at the same time an element of mystery. Although each step depends on the preceding ones, we are left in the dark as to how they were suggested to Archimedes. There is, in fact, much truth in the remark of Wallis to the effect that he seems 'as it were of set purpose to have covered up the traces of his investigation, as if he had grudged posterity the secret of his method of inquiry while he wished to extort from them assent to his results.'"[10] Wallis also said, "Not only Archimedes but nearly all the ancients so hid from posterity their method of Analysis (though it is clear that they had one) that more modern mathematicians found it easier to invent a new Analysis than to seek out the old."[11]

The reticence and apparent secretiveness of science are to be explained on other and better assumptions than that of jealously guarding a secret. The hierophantic tone of Heraclitus, who was supposed to be intentionally obscure, is only more pronounced than that of other writers of antiquity. He was more interested in uttering his thoughts than in justifying them. Not only the man in the street, but the scholar in his closet and the scientist in his laboratory also are most concerned about the result, the object they had primarily in view from the start; the steps, or means, by which one obtains that result are regarded as valuable only to that end and may be later disregarded as having at most an antiquarian interest. Not every one has the vivid sense of his intellectual experiences possessed by the late Dr. W. T. Harris, who could date precisely a partciular "insight." For most men the moment of supreme happiness comes when their several tracks of thought meet in a common center, forming a system. Henceforth, whatever wanderings may have gone before, they see the world

[10] Heath, *Archimedes*, 30 f.
[11] Quoted by Heath, *A History of Greek Mathematics*, II. 21.

contentedly from that point of view, and in their eyes the several parts of the system mutually explain one another. It is as difficult to gather from a philosopher's argument the history of his thinking as to disengage the real motive of an act from the reasons subsequently offered in justification of it; perhaps the actor is himself the poorest judge of his own motives.

One more stricture on Greek scientists calls for notice. One complains that they were too ambitious, trying to compass all nature, or at least an entire science, in their treatises. This is of course eminently true of Aristotle, who has remained the ideal of the Summists to this day; but it is not equally true of all Greek men of science, many of whom confined themselves resolutely within more modest limits. With this criticism Lord Bacon, who was hardly entitled to make it, combined the other that Aristotle never mentioned an author except to refute him.   We need not concern ourselves with the question how far these objections are justified, because our present aim is neither to praise nor to blame the pioneers but, so far as we may, to understand them.   Regarding the scope of his work, Aristotle, to mention him in particular, has been justly admired for that very reason, as being one of the few who have embraced the whole world in their philosophy.   Comprehensive as was its purview, his task was, however, vastly simpler than it would be if he were living now—the survey of even a single science is hardly within the reach of an individual today.   In his time knowledge, though extensive enough to tax the capacity of the greatest intellect, was not parcelled out in so many masses that it was impossible for one to master at least the leading principles of all; and since it was the natural history of the world that science aspired to tell, it was natural that it should be conceived as universal history.   It was of course inevitable that the student should be more at home in one part of this world than in another, and, except as the fundamental principles directly applied, it was equally inevitable that the contribution of the individual should lie in special fields. One must not infer that the Greeks were unfamiliar with the monograph—there were, in fact, great numbers of such essays on particular subjects; but it is true that they were not as a rule limited to the writer's own discoveries. Here the artist's desire to present a whole rather than a fragment asserted itself. This fact in turn goes far toward explaining why Greek writers in general rarely cite their sources. The practice of doing so really belongs to the literature of learning, which is the work of commentators and grammarians. For the rest, property rights in matters of knowledge were not recognized; to publish a work was literally to make its contents *publici juris*, common property. Hence there was no impropriety in using for one's own purposes any facts or theories which one found serviceable. Only where the author had given his work, as in verse, an artistic form, had he established a claim that decency required one to respect. Consequently there was little occasion to mention predecessors except where one felt the need of making clear the relation of ideas. If one require a modern parallel, let him consider the practice in making text-books. The greater part of the matter presented is likely to be practically the same in all that cover a given field, because it belongs to the tradition. The original contribution of the several authors is usually slight, and may often consist largely in a novel arrangement. Such jealous claims of priority of discovery and guarding against violation of copyright as we know today did not and could not exist—they are the product of a commercial age. The Greeks displayed considerable interest in discoveries and inventions from at least the Fifth Century onward, but the determination of the "sole first inventor" evidently gave them as much trouble as we experience in our time.

The more in detail one considers science, as it appears among the Greeks, the plainer it becomes that little has changed in all the succeeding centuries except in what are after all secondary matters. This is not said with any thought of disparaging the advances in methods and results, which it were foolish to minimize or deny, but rather because of the reasoned conviction that the first, determining steps in all directions were taken at the beginning. One of the most intelligent and critical historians of science has said,[12] "With Hippocrates closes the third period of the history of medicine, a period essentially constitutive not only for medicine but for all the other branches of intellectual culture. It is a decisive period in the destiny of the human race. All the germs of knowledge of the succeeding centuries are contained in it; henceforth everything will proceed from it. It is not a renascence, as in the times of Charlemagne, of Leo X, of Louis XIV; it is the spontaneous movement of the Greek genius, expanding in every direction and creating the best models and the most perfect types in every kind. This primordial fecundity, which has never recurred with equal potency at any time, has never been halted, so, descending from age to age, our Nineteenth Century is the legitimate offspring of the great age of Pericles. That Fifth Century of the ancient era is in the intellectual sphere what the first age of the world is in the material." What Daremberg says of the glorious spring-time of Greek thought will readily be recognized as true if one does not limit one's survey to the Fifth Century, where in certain special respects the germs are still more or less latent, but takes in the harvest-season of the later centuries, when the ripe fruition was abundantly garnered.

[12] Daremberg, *Histoire des Sciences Medicales*, I. 145 f.

One does well always to bear in mind the fact that science, as we understand it, is only a specialized employment of the same faculties that serve in the ordinary practical affairs of life.   Like every specialty, it is an offshoot of the commonplace; but, as is true of every interest that constitutes itself an institution, it does not merely continue the practices of the world at large, but adapts them to its own needs.   In time, it creates a ritual and a technique of its own, though the technique will still bear the impress of the special interest to which it owes its origin.   Some one has said that an institution is only the shadow of a man; it were better to say that it is the independent projection of an interest that was at least at the beginning vital and absorbing.   If that be true, one may fairly judge how strong was the intellectual impulse among the Greeks in the age that gave birth to science, especially as it shows at first so little inclination to serve practical ends.   One sees how even the terms most used in science were borrowed from practical life.   Thus the term investigation comes to us through the Latin from the Greek terminology of the hunt, which had already found metaphorical application in the courts of justice, as tracking a crime or a criminal.[13]   To the same source we owe the term *method*, which is only the systematized search.   In the same way one may find, for example, in the *Oedipus Tyrannus*[14] of Sophocles, where one sees a

[13] Plato was particularly fond of the figure because it lent animation to the dialogue; cf. *Phaedrus*, 270 c, 276 d; *Theaetetus*, 187 e; *Politicus*, 290 d; *Sophist*, 218 d, 235 d; *Parmenides*, 128 c; *Phaedo*, 79 e; *Republic*, 401 c., 432 c–d; Sophocles, *Oedipus Tyrannus*, 109, 221, 475.

[14] 158, 771.   The word is *elpis*, commonly, but erroneously, rendered "hope."   I discussed the notion, "On Certain Fragments of the Pre-Socratics," *Proceedings of the American Academy of Arts and Sciences*, XLVIII (1913), 696 f., in connection with Heraclitus, fr. 18, Diels.   To the examples there cited add: Herodotus, II. 2, 11, 26, 43, 120, III. 119; Hippocrates, *De Morbis*, IV. 35 (VII. 548 L.), *De Habitu Decenti*, 11 (IX. 238 L.), *Prognosti-*

relentless prosecution of a quest, the use of a word, borrowed from common speech, to express a theory, in the sense of a working hypothesis to direct research—it is no other than "expectation," "surmise." It is hardly necessary to insist on the influence of court-procedure in developing the technique for the finding and testing of truth, as one recognizes the importance of the law of evidence in court-proceedings. Of course, the technique varies with the nature of the subject under investigation, growing more and more refined as problems become more complex and the possibilities of error are detected; but the guiding principles remain essentially unchanged. All the important methods employed in science were represented in the practice of the law; observation, by the testimony of witnesses; analogy, by the argument from probability, which played so conspicuous a part in Greek forensic oratory; deduction, by the inferences from admitted facts and principles; and, finally, experiment, by the testimony of slaves under torture. To our minds the practice last mentioned appears to be neither humane nor likely to yield the certain truth, which alone might perhaps under certain circumstances justify it;[15] but it was looked upon as exemplifying the process of extracting the truth from unwilling witnesses. One of the Hippocratic writers obviously had this procedure in mind; for, after speaking of the symptoms of the patient which the physician must carefully note, he goes on to say,[16] "But when Nature will not willingly and of her own accord yield these indications, the science of medicine has discovered

---

cum, 15 (II. 164 L.), 19 (II. 164, 168 L.); Galen, De Placitis Hippocratis et Platonis, I. 165, ed. Müller; Plato, Republic, 517 b.

[15] One may question whether it is more cruel than the so-called "third degree" to which suspects are now at times subjected.

[16] De Arte, 12 (VI. 24 L.). Theodor Gomperz, Die Apologie der Heilkunst², 140 f.

means of constraint, whereby Nature, without damage to herself, yields them under compulsion, and by displaying them discloses to those who are versed in the science, what they are to do." Sophocles in his *Oedipus Tyrannus*[17] makes the King reply to the chorus, who suggested that, since Apollo had imposed the duty of expelling the slayer of Laius, it was for that god to discover him, that the suggestion was good but impracticable, since there were no means by which the gods could be forced to declare what they chose to hide. The figure and its application to scientific experimentation were taken over by Bacon and have received their classical expression in Goethe's *Faust*:

> Geheimnisvoll am lichten Tag
> Lässt sich Natur des Schleiers nicht berauben,
> Und was sie deinem Geist nicht offenbaren mag,
> Das zwingst du ihr nicht ab mit Hebeln und mit Schrauben.

The foregoing is introductory, written in the hope that it may be helpful to the reader in his effort to understand and appreciate the work of the pioneers in science. He may be impatient to have done with generalities and proceed to particulars which may enable him to judge for himself whether the ancients actually exhibit the ideals and methods which we attribute to them. If such be his desire, the only adequate way to satisfy it would be to make a thorough study of the scientific and philosophical writing of the Greeks and, in fact, of the entire body of their serious literature, because it is often in less technical works that one most clearly follows the processes of their thought. Nothing would give me so great satisfaction as the thought that this general discussion had in some way served to enlist the interest of even a few scholars who would devote themselves to the subject; for I know of nothing better

[17] 280 f.

calculated to open one's mind and help one to take a view at once comprehensive and critical of one's own special field of investigation than a broad historical survey of the work that has been done in it—the wider the survey, the more likely is one to discover its direction, its limitations, its possibilities, and its immediate problems. One might cite as examples men like Sir William Osler and others, who saw their own work in truer perspective because they took the larger view. The general reader, even the generality of readers devoted to science, can not, of course, be expected to undertake a study of Greek literature as a whole; and it would ill serve the purposes of this book either to refer generally without a bill of particulars to what the pioneers did, or to give extensive selections from their writings exemplifying the different processes.

It is, then, a question of selecting here and there passages which illustrate the way in which the Greeks attacked their problems. The task is a difficult one, and in the nature of the case the result is bound to be not wholly satisfactory in itself, and, besides, to invite unfavorable criticism, because no two scholars would naturally agree on the examples to be chosen. Disregarding the latter aspect of the task, as in any case inevitable, one must nevertheless insist on the difficulty of representing the processes of science by selections, which grows out of the nature of the scientific procedure itself. After all, as has already been said, science is only a more or less technical employment of the intellect, and the mind does not proceed by separate and distinct acts to its goal. Even if it did so, we should still be little advantaged by that fact, since, as we have tried to show, the form adopted in exposition, on which one must depend for illustration, is not the order of thought.

In order to exhibit the several operations involved in the work of science one must make at least a rough analysis of the process.   Let us say, then, that it consists primarily in framing and in testing generalizations.   If one assume that the scientist begins with a *tabula rasa*, the first data are sensations apparently unconnected; these sensations the mind groups and fashions into "objects," which are progressively defined, related, compared, and contrasted, and on the basis of similarities and differences classified according to some scheme or principle.   Thus stated, the process seems simplicity itself.   Unfortunately it is not in fact so simple, for the sensations are not unconnected, nor are the objects into which they combine mere data—the noting of sensations and the grouping of them into objects are alike due to the directing influence of the same interests which in the end determine the classification.   Consequently the classification is virtually given with the objects that are combined in it and even with the sensations which the selecting mind has chosen to take notice of.   Of course the naïve thinker is not aware of this fact, but regards the resulting world, which he accepts as wholly objective, as something which he must master.   His task is the easier because it is half accomplished when he consciously takes it in hand; for in his scheme of classification he has more or less definitely anticipated his most fundamental generalizations.   As for the process of checking his conclusions, that too is already considerably advanced; for in the act of framing his system he has exercised the function of rejection by noting differences, and on the basis of positive marks, which unfit an object for inclusion in one group, he has erected other groups obeying other principles.   If science were content to be merely descriptive, we might say that in principle its task was accomplished when it had achieved a satisfactory scheme into which it could include every phenomenon observed in the world.   The process, which in this general statement seems so simple, is really extremely complex. The several operations of defining, relating, classifying, etc., do not in general succeed one another in this or any other definite order, but combine in different ways and mutually condition one another; one and the same operation having in fact several functions, one or other of which will be more prominent in thought according to the momentary interest. Any attempt to illustrate the several operations separately will inevitably disrupt the movement of thought, and consequently result in something like a caricature. Moreover, there is another fact to be considered. If man were governed by one sole interest, or if a single interest were so strong as to dominate the lesser ones and keep them permanently in subjection, there would probably be no great difficulty. The selective principle which shaped his first objects and classifications would suffice to keep them fixed, and all would be harmonious. But, whether for better or for worse, man is not so constituted. Different interests rule his thought and actions at different times, and each creates its own world. And each world, with its objects and generalizations, constitutes a problem for every other world.

This problem becomes acute because of our desire to have one world instead of many. Science insists on something more than mere description, which would be possible even if there were only a series of unrelated pictures. In demanding a cause for everything, and a hierarchy of causes culminating in a supreme cause, which shall account for every thing, for all our particular worlds, science is inexorable and creates a situation which brings out all the latent anarchy of our thinking. This is in fact the special rôle of science: it has apparently come not to send peace, but the sword. It is in the conflict between the different worlds of our creating that it is most at home; and the things it is most forward in challenging are the objects (facts) and generalizations that are the products of less universal interests. One may compare this conflict to that between the emperors and the petty barons. In judging Greek thought it is well to observe that in the effort to bring order into the intellectual world it is as uncommon today as it was in ancient times to go back to first principles. This is the result of the separation of science from philosophy, to which one relegates the ultimate problems, though science too often assumes that its analysis is final.

Since science, as has been repeatedly stated, does not differ essentially from any other rational employment of the mind, it would be futile to attempt to illustrate every phase of its procedure; all that one may reasonably require is that it be considered in its chief aspects. Among these we shall single out for special treatment observation, classification, analogy, and experiment; and we shall want to see how far they are recognized in theory as well as in practice.

II. OBSERVATION AND INDUCTION

Observation is an ambiguous term; or rather, it has, in the sense of mere perception, no special significance. As a matter of fact, one may question whether in the developed human mind it ever really stands alone. No one who is at all acquainted with Greek civilization doubts that those who created it had very keen senses; but that is at least equally true of the normal savage. One might even argue that civilization tends rather to blunt than to sharpen the senses; and in the most primitive men observation passes directly over into inference. We use "observation" and "seeing" not only of physical but also of mental operations, including recognition and understanding, which involve additional activities of the intellect. As seeing is not entirely passive, it is merely a phase of a larger operation directed by an interest. All this is so elementary that one need not insist upon it. But, recognizing observation as purposive, we must add that it accompanies the entire mental process. Regarded as furnishing the bare materials of which the mind builds its elaborate structures, it is not greatly esteemed either by the individual or by science. Its value is chiefly appreciated at a later stage, when question arises whether the structure built upon it will stand or not; in other words, observation plays its principal rôle in testing the inferences which have been drawn from the preliminary survey.

It is well, however, to remind ourselves that the conclusion inferred from observation need not be an abstract theory, but in practical life more frequently assumes the form of a course of action. The astonishingly acute and accurate observations of the Hippocratics, which the most competent historians of medicine never weary of lauding, served, and were intended to serve, practice rather than speculation. They sought, as some modern physicians seek, to form a total mental picture of the case they were studying, and that total picture embodied their clinical observations. We find interesting directions for this procedure, which show that its complexity did not escape their attention: "Sum up the origin and beginning of the disease from as much talk as possible with the patient and from the facts gleaned little by little, bringing them all together and learning whether the symptoms bear a family resemblance, so that a single picture may be framed of their differences.[1] That is the proper method; in such wise you will confirm what is proper and disprove what is not."[2]

We are, however, more interested in observation in the service of theory. It is sometimes said that Greek thinkers did not base their conclusions on observation, but, as it were, shut their eyes to the world and speculated on the basis of pre-conceived notions. That view is of course absurd; for whence came the pre-conceived idea? The truth is that they did observe, and did speculate; and perhaps they actually rendered their greatest service to science by their speculations. This will perhaps be regarded by scientists as a hard saying; but may it not be true? Man had presumably been observing ever since he appeared on the earth; what differentiates the earliest Greek pioneers from their predecessors is that they sought rational explanations for what they observed. As has already been said, the preliminary survey was cursory—often quite superficial. The real test and significance of observation lies in its use as offering confirmation or disproof of theories. Aristotle, who has been severely criticised for his failure to base his

[1] Cf. Aristotle, *Topica*, VIII. 18, 108 a 38 ff.
[2] Hippocrates, *Epidem.*, VI. 3, 12 (V. 298 L.).

speculations on observed facts, repeatedly urges the necessity of making theories harmonize with the facts;[3] and after a most interesting discussion of the generation of bees, which contains much admirable criticism of other theories, along with some false generalizations of his own, he says,[4] "Such appears to be the truth about the generation of bees, judging from theory and what are believed to be the facts about them; the facts, however, have not yet been sufficiently determined; if ever they are, the credit must be given rather to observation than to theories, and to theories only if what they affirm agrees with the observed facts." In certain fields, such as astronomy, observations of great accuracy were made, and became the foundations of theories which have been proved to be true. Of more exact observations in optics and mechanics we shall have to speak later.

Despite everything that one may say in favor of Greek science, it remains true that compared with the best modern practice the procedure of the earlier Greeks was superficial and hasty. The data which call for explanation were not collected and scrutinized with the care demanded in our time. In part the failure to do so is chargeable to the want of instruments and techniques necessary to the closest observation and the finest discrimination; but in far greater measure it was due to other causes. The newly developed arts of rhetoric and dialectic, which in the hands of the Sophists had proved powerful weapons for offense and defense, and the new-found speculative faculty, which in dealing with abstractions had demonstrated a capacity for discovering truth without resort to the data of sense, had in fact produced a state of mind little inclined to the drudgery of collecting and sifting facts, except in lines of

[3] *E. g. De Motu Animal.*, 1, 698 a 12 ff.
[4] *De Generatione Animal.*, III. 10, 760 b 27 ff.

inquiry, such as medicine, where the practice had been long
established.

Theory and practice do not always progress *pari passu.*
It may be useful nevertheless briefly to consider Aristotle's
theory of induction.   In an earlier chapter[5] we saw how he
conceived of the development of science out of crude sense-
perceptions.   In his logical treatises[6] he distinctly recognizes that the method of induction is nothing more than the
conscious and intentional practice of the procedure by
which in ordinary affairs we pass by abstraction from the
particular to the universal.   The particular, however, is
given only in sensation; hence induction is wholly dependent
on it.[7]   Moreover, the universal is implicit in the particulars.[8]   Induction depends on classification,[9] and classification in turn is based on observed likenesses and differences.
"Examination of likenesses," we are told,[10] "is useful with
a view both to inductive arguments and to hypothetical
reasonings . . . . because it is by means of an induction of
individuals in cases that are alike that we claim to bring
the universal in evidence, for it is not easy to do this if we
do not know the points of likeness.[11]   It is useful for hypothetical reasonings because it is a general opinion that
among similars what is true of one is true also of the rest."
Since, according to Aristotle, though the universal is in
itself more knowable, the particular—the object of perception—is to most people better known, especially in speaking to a crowd one should proceed from individual cases,

----

[5] P. 44.

[6] *Analyt. Post.*, II. 19, 100 b 3 ff.

[7] *Ibid.*, I. 18, 81 a 38 ff.

[8] *Ibid.*, I. 1, 71 a 5 ff.

[9] *Topica*, VIII. 2, 153 a 21 ff.

[10] *Topica*, I. 18, 108 b 7 ff.

[11] For the principle of concomitant variations, see *Topica*, II. 10, 115 a 3 ff.

from the known to the general and to the unknown, although deductive reasonings are more compelling and effective against disputatious people.[12]   Ideally, of course, induction proceeds by an examination of all the cases;[13] but in practice, especially in argument, one may be content with less than a complete survey, unless one either finds a negative instance or has one presented in objection to the generalization.[14]   The several sciences, as has already been pointed out, derive their supreme principles by induction.   "I mean for example, that astronomical experience (observation) supplies the principles of astronomical science; for once the phenomena were adequately apprehended, the demonstrations of astronomy were discovered.   Similarly with any other art or science.   Consequently, if the attributes of the thing are apprehended, our business will then be to exhibit readily the demonstrations.   For if none of the true attributes of things had been omitted in the investigation, we should be able to discover the proof and demonstrate everything that admitted of proof, and to make that clear whose nature does not admit of proof."[15]   The demonstration here meant is that given by deductive reasoning, and the propositions which do not admit of proof are the highest principles, which can not be derived from more general ones.

So far as general statement goes, this account of induction is reasonably satisfactory; for Aristotle was well aware that induction can not in itself yield a true universal or lead to absolute certainty.   Where the theory falls short of perfection is in the directions for making and testing the inferences to which induction points.   Aristotle tells us

[12] *Topica*, I. 12, 105 a 13 ff., VIII. 1, 156 a ff.; *Analyt. Priora*, II. 23, 68 b 35 ff.
[13] *Analyt. Priora*, II. 23, 68 b 29 f.
[14] *Topica*, VIII. 2, 157 a 34 ff.; cf. Mill, *Logic*, Bk. III, Ch. III, §2.
[15] *Analyt. Priora*, I. 30, 46 a 19 ff.

that induction is valuable because it leads to the knowledge of the cause;[16] but he means something quite different from the proximate causes which science now seeks—for it is the formal cause or essence, which is practically identical with the universal, that he has in mind. The source of his difficulty lies ultimately in the attitude toward knowledge which he shared with Plato. Though both philosophers recognized the need of relating the universal to the particular, neither could grant the highest degree of truth or evidence to the latter. While Plato thought of concrete things as mere reminders of the ideas, and the world as something, like the flesh and the devil, to be renounced, Aristotle, though ostensibly claiming for things of sense a prime reality, could find true knowledge and certainty only in deduction, or reasoning from universals, which had been indicated rather than proved by induction. If in theory induction thus rested on assumptions which foredoomed it to be ineffectual, it is not surprising that in practice it should not have yielded better results. The remarks above quoted about astronomy will illustrate the Aristotelian point of view, which hardly differs in principle from the Platonic. The principles are the chief concern of the philosopher; once they are attained—it little matters how, since even on Aristotle's view induction does not yield certainty—the consequences in detail are to be deductively demonstrated. That there is here a complete break-down in the theory is clear; but it was inevitable so long as the practical technique of investigation remained undeveloped.[17] While modern theories of induction labor under essentially the same disability as the Aristotelian, since they also fail to establish the claim of induction to attain complete certainty, scientific procedure, by the constant interplay of in-

16 *Analyt. Post.*, I. 31, 88 a 2 ff.
17 Cf. Zeller, *Die Philosophie der Griechen,*[3] II. ii, 245 f.

duction and deduction, of observation and the testing in detail of inferences drawn from it, has in fact gone far toward establishing the legitimacy of its method. Seeing that deduction yields truth only so far as it is contained in the generalizations with which it starts, and that induction furnishes these generalizations, the truth of science depends on the functioning together of particulars and universals in the total process, rather than upon any single factor in it.

As we might expect under the circumstances then existing, science scored its most notable successes in fields where observation played a minor rôle. One thinks, for example, of the theory of the geographical zones, which is, on doubtful evidence, attributed to Parmenides. It rests on the assumption of a spheroidal earth, situated at the center of a spherical universe. The celestial zones were easily marked by those who observed the seasonal changes in the appearance of the stars; it was only necessary to apply the most elementary principles of spherics, which had been well advanced by Eudoxus, drawing lines from the center of the earth to the points where the planes of the several celestial zones cut the periphery, and the terrestrial zones could be traced on any globe. In theory it was at once obvious how to determine the location of the zones on the earth's surface; actually the observations necessary for their determination had to wait until the time of Dicaearchus and Eratosthenes, though some rough approximation must have been known to Aristotle when he estimated the circumference of the earth. It happens that it makes no practical difference that the earth is not actually at the center. Moreover, it seems fairly clear that the discovery that the earth is a globe was due to purely abstract mathematical speculation —once the hypothesis was suggested, the familiar proofs were found. It is characteristic of the time and the point of view which then obtained that Plato is said to have suggested the problem which led directly to the hypothesis of the rotation of the earth on its axis and mediately led Aristarchus of Samos to propose the heliocentric theory. A certain modicum of observation was of course necessary to the definition of the problem; but the solution seems to have been almost entirely a matter of mathematical theory. One suspects that it was precisely the deficiency of detailed observation which made it possible for most astronomers to reject the latter theory until Copernicus revived it. That it was in fact primarily, if not entirely, a mathematical solution is made highly probable by the alternative theory, propounded by Eudoxus. Sir Thomas Heath says,[18] "It is the theoretical side of Eudoxus's astronomy rather than the observational that has importance for us; and, indeed, no more ingenious and attractive hypothesis than that of Eudoxus's system of concentric spheres has ever been put forward to account for the apparent motions of the sun, moon, and planets. It was the first attempt at a purely mathematical theory of astronomy, and, with the great and immortal contributions which he made to geometry, puts him in the very first rank of mathematicians of all time. He was a man of science if there ever was one."

One may at first incline to discount this statement as coming from a mathematician; but reflection will lead one to accept it at its full value. Though the theory of Eudoxus was not founded primarily on minute observation, and though it has had to be discarded, it is noteworthy not merely for the ingenuity it displays, but also for the observations which it directly occasioned. It matters little whether one begins or ends with observation; it is, of course, the final test, but theory alone can give meaning and direction to inquiry, and one must judge the individual scientist

[18] *A History of Greek Mathematics*, I. 323.

rather by the impetus he gives than by the formula that he happens to propose.   In process of time the need of long and accurate observation in astronomy became increasingly clear.[19]

A good illustration of the procedure of the mathematicians is provided by Archimedes.   Mention has already been made of Analysis, as practiced by them.   It consists in assuming the truth of a proposition which can not be directly established, and working back from that to known equations; if the deduction, or Synthesis, succeeds in reaching such equations, the required proof is found.   This is in fact the exact counterpart of the procedure followed in proposing a working hypothesis and then subjecting it to rigorous tests.   In his *Method* Archimedes describes his own way of finding the solution of certain problems, such as the center of gravity of various figure.   To do so by asbtract reasoning without any indication of the precise locus would be difficult; but by substituting for the abstract figure, say, a thin material plate of the required form and balancing it upon a sharp point, it becomes possible empirically to determine the location with approximate precision, and then by strict mathematics to demonstrate what the tentative method had suggested.   No one will dispute either the ingenuity or the utility of such a procedure, much less deny its scientific character, though it does not rest on elaborate induction.   It is a short-cut, nothing more; but it proved to be, in the hands of Archimedes, a heuristic instrument of the first order.   Actually the method of Archimedes, as he sets it forth, is of course not so crude as that.   Though his constant reference to the beam of a balance suggests that in the simpler figures he may at first have resorted to such means, in the treatment of spheric and conic sections, and of cylinders and prisms, with which he is chiefly concerned,

[19] Seneca, *Nat. Quaest.*, VII. 3, 1.

his reasoning is essentially geometric, except that he obviously imagined actual masses instead of ideal figures.
Brilliant mathematician though he was, the approach to
pure mathematics by way of mechanics was, as he confesses,
of great service to him.   Archimedes was, moveover, keen
enough to recognize the value of theorems, though they were
propounded without demonstration.   "Certain things," he
says,[20] "first became clear to me by a mechanical method,
although they had to be demonstrated by geometry after-
wards because their investigation by the said method did
not furnish an actual demonstration.   But it is of course
easier, when we have previously acquired, by the method,
some knowledge of the questions, to supply the proof than
it is to find it without any previous knowledge."   "This,"
he adds, "is a reason why, in the theorems that the volumes
of a cone and a pyramid are one-third of the volumes of the
cylinder and prism, respectively, having the same base and
equal height, the proofs of which Eudoxus was the first to
discover, no small share of the credit should be given to
Democritus, who was the first to state the fact, though
without proof!"

This procedure is, of course, most successful in mathematics; and it is not surprising that the Greeks, who felt most
at home in the world of ideas, advanced so much further in
this department than in the fields where extensive and minute observations were necessary.   Nevertheless, they were
blind neither to the world nor to the need of the close inspection of nature in detail.   We may now give a few illustra-

[20] *Archemedis Opera Omnia*, ed. Heiberg, vol. II, p. 428 f.; the translation
is that of Heath, *A History of Greek Mathematics*, II. 21.   Compare Plutarch, *Vita Marcelli*, XIV. 5, and *Quaest. Conviv.*, VIII. 2, 1, 7 for the procedure of Archytas and Eudoxus.   The whole treatise has been translated
into English by Lydia G. Robinson, *Monist*, vol. XIX (1909) and by Sir
Thomas L. Heath, *The Method of Archimedes*, Cambridge, 1912.

tions of their capacity in this direction.   Even the earliest
philosophers, who are sometimes supposed to have propounded theories without using their eyes, really based
them on observations.   Aristotle and Theophrastus, not
knowing why Thales asserted that all things sprung from
water, were right in suggesting observations which might
have led him to that conclusion; if Anaximenes said, as is
reported, that when air is most uniform it is invisible, but
becomes visible by heat, cold, moisture, and motion, it is
easy to guess the phenomena he noted and reflected on.
The fault of the philosophers, if fault it may be called, is
that they were too ready with explanations.   Not all
Greeks, however, were so forward with theories.[21]   This
is notably true of many of the early physicians.   Daremberg is quite right in saying,[22] "In the *Epidemics* [of Hippoc-
rates] etiology is at the stage of observation pure and
simple, and it is precisely this character that constitutes
the great merit of the work and makes it safe from all
attacks."   Elsewhere[23] he refers to the clinical record in the
Prognostic of the acute disease which so much engaged the
attention of the Greek physicians, and remarks that it was
so exact that Littré was able to identify it with certainty
as intermittent fever.   In fact, there is much in the Hippocratic treatises that is extraordinarily acute and accurate,
and which has been observed and verified only by the latest
and most careful investigations.   Littré[24] calls attention to
a striking instance: "The *Second Prorrhetic*, section 17,
recommends that when the throat fills with blood one
should see whether a leech has attached itself to the walls.

[21] For empirics among physicians, see Plato, *Legg.*, 859c; among lawyers,
ibid., 938 a.

[22] *Histoire des Sciences Médicales*, I. 111.

[23] *Ibid.*, I. 109.

[24] *Oeuvres Complètes d'Hippocrate*, IX. 3 f.

Ancient critics, casting doubt on the fact, wanted to interpret the Greek word in some other way than 'leech'—by
lesion, or some sort of ulcer.   But most careful observation
has proved that such an affliction caused by leeches imbibed
with water from a spring or pond is not unexampled.   The
Hippocratic writer had seen such a case and noted it."   It
is in the spirit of the Hippocratics that Thucydides, in
speaking of the devastating plague that visited Athens
shortly after the beginning of the Peloponnesian War, says,[25]
"Anyone, whether physician or layman may express his
opinion regarding its origin and the causes which he thinks
sufficient to account for so great a departure from normal
conditions; but I will describe its nature and the symptoms,
by noting which, if it should ever recur, one may with this
knowledge not fail to recognize it; for I had the disease myself and saw others suffering from it."   The description
which he gives of the disease, minute and clear as it is, has
not, unfortunately, enabled physicians to identify it with
certainty; which is not to be wondered at, because diseases
have a way of varying with varied conditions; but no reader
will fail to recognize the painstaking and trained observer.
The same characteristics mark the historian's observations
on the social, moral, and religious effects of the plague and
of the revolution at Corcyra.[26]   The latter account is a
masterpiece of observation, analysis, and inference which
has hardly, if ever, been surpassed.

From what was said above about Aristotle's debt to the
medical tradition, one would expect him to display at times
the same capacity for accurate observation, especially in
his biological treatises.   A few examples will suffice for
our present purpose.   "All viviparous animals," he says,[27]

[25] II. 48.
[26] III. 70–84.
[27] *Historia Animal.*, I. 9, 491 b 27 ff., tr. Thompson.

"have eyes, with the exception of the mole. And yet one might assert that, though the mole has not eyes in the full sense, yet it has eyes in a kind of way. For in point of absolute fact it can not see, and has no eyes visible externally; but when the outer skin is removed, it is found to have the place where eyes are usually situated, and the black parts of the eyes rightly situated, and all the place that is usually devoted on the outside to eyes; showing that the parts are stunted in development, and the skin allowed to grow over." This is actually true, we are assured, of the *Talpa caeca* of Southern Europe. Aristotle frequently recommends his readers to investigate for themselves, by dissecting the animals which he describes; indeed, there is abundant evidence of the practice of dissecting animals for scientific purposes from Alcmaeon of Crotona, in the Sixth Century, onwards. In speaking of the internal organs of certain testacea, Aristotle says,[28] "While there are some points which can be made clear by verbal descriptions, there are others which are more suited to ocular demonstration." It is plain that he had himself inspected, and in many cases dissected, most of the animals which he describes. Similarly Theophrastus was familiar with almost all the plants he discusses. How soon illustrations were added to herbals, can not perhaps be determined, but it was a regular practice in later antiquity and, in the beginning at least, the drawings were of course made from nature.

We naturally have a greater interest in observations of larger scope which form the basis of inference. A good example occurs in the Hippocratic treatise *On the Nature of Man*, where we read:[29] "Some diseases come from our regimen, others from the air we breath. We must distinguish between them in this way: when many persons

[28] *De Part. Animal.*, IV. 1, 680 a 1 ff., tr. Ogle.
[29] *De Natura Hominis*, 9 (IV. 52 f., L.).

are seized by one and the same disease at the same time, the cause must be sought in that which is most common and which we all most use. This is the air we breathe. For it is plain that it is not due to our several modes of life, when the disease attacks all alike, one after the other—young and old, men and women, wine-bibbers and teetotalers, eaters of barley-cake and eaters of bread, men at hard labor and idlers who do little—their manner of living could not be held responsible, when men of every mode of life are seized with the same disease. But when all sorts of diseases appear at the same time, it is plain that the cause lies in men's several ways of living, and one must counter the cause of the disease, as I have said elsewhere, and effect a change of regimen. For obviously the patient's usual manner of living is unsuitable, either altogether or for the most part, or in some particular; this one must determine, and effect a change, with constant regard to the patient's constitution, age, and complexion, and direct the treatment according to the season of the year and the character of the disease—adding here, subtracting there, countering by medicaments and diet the tendencies of the patient's time of life, of the seasons, of the complexions, and of the diseases.'' However, one may regard in detail the procedure here recommended, there can be no doubt about the writer's purpose to make sure that every aspect of the patient's condition shall be most carefully observed, as a necessary preliminary to the physician's prescription. Indeed, it is unmistakable that the intended inspection required tentative steps which must be of an essentially experimental character. There is, of course, likewise implied a certain amount of theory, but in the procedure suggested it does not at all obtrude.

As we have seen, Aristotle's works fall into two classes, one dominated by the speculative interest, which ruled in the school of Plato, the other showing the influence of the empirical methods, which were naturally cultivated in the medical schools.   One must not think of these influences as wholly opposed to one another; for it is clearly an exaggeration to regard Plato as the arch-enemy of observation, and the medical fraternity was far from being averse to speculation.   The human mind, except when it sets itself resolutely to conduct its operations according to a predetermined theory, does not work that way.   But there were in fact influences of the sort making themselves strongly felt in Aristotle's time, and they were not easy to bring together for harmonious cooperation.   They show themselves, if not directly opposed, at least for the most part separated in his several works; and the logic, in which he formulated the principles which governed the processes of thought, bears unmistakable evidence of his failure to coordinate them.   In his school there was a similar division among his pupils and successors, some displaying a predominant interest in the speculative, others in the observational and historical sides of his activities.   It is noteworthy that the former betrayed little originality, confining themselves for the most part to restating or at best developing his thought, while the latter, represented especially by Strato and Dicaearchus, contributed greatly to the advancement of particular sciences.   Strato was in fact one of the foremost scientists of Greece, and there are few losses which we have more reason to regret than that of his works. Whether he actually had a connection with the Ptolemies or not, it is not easy to decide; but it seems certain that the Lyceum under his presidency exerted a powerful influence on the institutions for research which the Ptolemies founded in Alexandria.   Directly and indirectly these institutions, where many of the foremost scholars and scientists of Greece were brought together and liberally provided with the necessary facilities for work, account for a good part of the achievements in science which made the last centuries of ancient Greek thought so notable.

We shall have to recur to this subject later in treating of experimentation. Our present concern is with the development of the empirical method. To trace its history in detail, even if our sources sufficed for that purpose, as they actually do not, would require so involved a discussion that it would be difficult to follow the train of thought. We shall, therefore, attempt a brief sketch only, in the hope of bringing out clearly the main outlines of the story. Owing to the fragmentary state of our sources of information a certain amount of speculation is unavoidable in any attempt to reconstruct the history of so complicated a movement, but we shall try to confine it within the narrowest limits.

The earliest philosophers, as we have seen, worked in the spirit of the scientist. If the range of their observations was insufficient to justify the theories based upon them, they may be pardoned because experience had not yet enforced the lessons of caution; for they were dealing for the most part with matters in which mistaken judgments are not immediately visited with dire consequences. In medicine it is quite a different matter. Moreover, the vague and general hypotheses of the physicians, when they were propounded, were fortunately so far removed from the activities with which they were concerned that they presumably had little direct effect on the course pursued with the sick —there it was a question of definite prescriptions, which were likely to be suggested by success in similar cases. Consequently medical practice must always have been in good part empirical, no matter what theories the practitioner might profess. In fact, as we have seen, the prevailing spirit of the medical fraternity, especially of the Old School, called for the closest observation of the patient and

of the conditions which might affect him.   How great was
the influence of medicine in the Fifth Century we are in no
position to determine, although it is clear from Plato, Xenophon, and Aristotle that it shared with mathematics the
honor of being regarded as the typical example of scientific
procedure.   We know of a number of centers in which medical schools flourished; but our information regarding some
of these schools is so meager that we can not go beyond
conjecture in tracing their possible influence.   We know
most about those of Cos and Cnidus; but even there we
are too ill informed to determine with certainty regarding
some of the works contained in the Hippocratic collection
whether they are to be assigned to the one or the other.
Somewhat clearer is the mark of the school of Crotona in
Magna Graecia, which gave rise to the Sicilian School, and
later to the Attic.   Others, like that of Cyrene, are known
to us practically only by name.

There was another school of medicine, however, which,
though we actually know next to nothing about it, would
appear to have exerted a considerable influence—that of
Abdera.   This city, situated in Thrace, had early connections with Ionia, with Clazomenae, Teos, and Miletus.
The scientific interests first fostered in Ionia evidently there
found eager followers, as witness Protagoras, the foremost
of the Sophists of the Fifth Century, Democritus, and perhaps Leucippus, exponents of the Atomic theory; and it is
not improbable that Nicomachus, the father of Aristotle,
pursued his medical studies at Abdera.[30]   At all events, it
is noteworthy that in all this group, including Aristotle,
there is displayed a strong leaning toward sensationalism
in the theory of knowledge, which received its first comprehensive expression in the teachings of Protagoras.   Upon

[30] Supposing that to be the case, we should be able to account for Aristotle's great interest in the Atomists.

him depend all later adherents of the doctrine, including, beside the Atomists and their successors, the Epicureans, the Stoics, and the Sceptics. How far the logical theory of induction, which rests upon the acceptance of the senses as the final resort in the determination of fact, had been carried by Aristotle we have already attempted to show. It was pointed out that this movement of the Stagirite's thought was apparently due to the influence of the medical tradition, with which he must have been made acquainted by association with his father, and the limitations of his procedure in this respect were noted. These limitations were in fact those which characterized the practice of the physicians of the Fifth and early Fourth Centuries. We have now to call attention to the important part which the later medical schools played in the further development of the theory of induction.

In the century after Aristotle, the movement away from intellectualistic dogmatism gathered strength. Among his successors in the Lyceum, Strato displayed the strongest bent for observation and experiment, but we have no record of any alterations or additions he may have made in logical theory. This is not surprising, because theory in general follows rather than precedes practice. The other notable experimenters of his generation, the foremost investigators in the fields of physiology and anatomy, Herophilus and Erasistratus, likewise seem to have contented themselves with the actual exploration of the body, in which respect their influence was great, leading to the formation of schools which continued to exist at least until the Third Century, A. D. Erasistratus, as we have seen, beautifully expressed the fascination of scientific research; but if he laid down rules for investigation beyond the requirement that one must base theory on observation or test it by the same, we have no knowledge of them. We shall later give examples of his experiments. As for Pyrrho, the founder of the Sceptic school, his attitude of completely suspending judgment did indeed challenge dogmatism, but it could contribute little in a positive way to the advancement of science. The Stoics and Epicureans based their theories of knowledge entirely on the data of sense, materialistically interpreted; but neither school at first advanced views calculated to assist science. We shall, however, presently find a late Epicurean who made a considerable addition to the Aristotelian doctrine of inductive inference; and Epicurus himself, who is usually regarded as having slighted logic (he called it "canonic"), seems to have understood a principle which even Bacon did not grasp. In the *Epistle to Pythocles*,[31] which, if not actually written by him, certainly contains his doctrines, a distinction is expressly drawn between phenomena which admit of but one explanation, and phenomena which admit of several. Again and again the latter class is mentioned by him and by Lucretius. Just how it was to be determined to which of these classes a given phenomenon belonged, is not made quite clear; though of course the procedure must have been the familiar one known to logicians as screwing together cause and effect. But one can not fail to see that Epicurus was well aware of the principle of the plurality of causes.[32]

It was in the Second and First Centuries B. C. that theory finally overtook practice, which had now become among scientists predominantly empirical. The degrees of probability were carefully studied, as well as the forms of the

[31] Diogenes Laërtius, X. 86–87.
[32] Mill, *Logic*, Bk. V, *Ch*. III, §7, says that the ancients did not recognize this principle. It may be true that the principle was not expressly formulated as such, but Aristotle several times (*e.g.*, *Poet.*, 1460 a 19 ff.; *Soph. Elench.*, 167 b 1 ff.) pointedly remarked upon the fallacy of inferring causes from effects.

inductive syllogism which may lay claim to cogency. So
long as one entertains no doubts regarding the possibility
of attaining certainty, probability may be neglected, as of
minor importance; but when criticism issues in scepticism,
one may welcome even that poor surrogate.   It was in this
spirit that the earlier Sceptics, Pyrrho and Arcesilaus,
treated probability; accepting it as a basis for conduct in
default of the complete assurance, the possibility of which
they virtually denied.   Carneades, head of the New Academy in the Second Century B. C., while rejecting the claim
of assured certainty, put forward by the dogmatists, sought
nevertheless to secure a reasonable foothold for the conduct
of life.   He could not, of course, acknowledge the report
of the senses as certain; but, though our organs were not
accurate, he thought they might be allowed to be sound, or
healthy.   Sensation has a two-fold reference: so far as it
purports to report what exists objectively, we have no
means of testing it, but in relation to ourselves it may command greater or less credence according as it approves itself
as giving a mere appearance of truth or a clear verisimili-
tude.   Credence will accordingly be of different grades.
The lowest will arise when the sensation itself makes the
distinct impression of truth, without standing in a definite
relation to others; the next stage is reached when the sensation is confirmed by all other relevant sensations; the
third and supreme degree, when an examination of all the
relevant sensations yields a mutual confirmation.   The
first degree is persuasive; the second, persuasive and immovable; the third, persuasive, immovable, and thoroughly
reconnoitred or inspected.   Thus, in the epistemological
theory of Carneades, which for him had importance primarily in the domain of ethics, we have the exact counterpart of the scientific process of erecting working hypotheses
and testing them by the most meticulous scrutiny.   While

Epicurus, for example, had contented himself with saying that an opinion was true if observation confirmed it and brought out no negative instance, the theory of Carneades requires in addition a systematic exploration of all relevant data of sense. This development undoubtedly reflects the improved experimental method practiced in the Alexandrian schools. The Academics, of course, could not concede that the results so obtained were more than probable; but they were at least willing to act upon the assumption that they might depend upon them. Favorinus, however, while generally avoiding the Stoic terminology, as implying too certain apprehension of truth, did concede that objects ascertained in the manner we have described were "known."[33]

As to the inductive syllogism, Aristotle had pointed out that it is strictly valid only when it proceeds through an enumeration of all the cases, in other words, when $A = B$.[34] But since a knowledge of all the particulars is generally impracticable, inductive reasoning must be inconclusive; we are, therefore, bound to have recourse to probable inference. It was here, as we have already noted, that Aristotle failed to perfect his theory of induction, and the problem which he left half-solved was taken up at this point in the last centuries before the Chrisitan era. We possess a considerable fragment of a treatise of Philodemus *On Inductive Inferences.*[35] which was found at Herculaneum. Philodemus himself contributes little to the matter in hand, practically confining himself to a report of the doctrine of two Epicureans, Demetrius and Zeno of Sidon, who date from the beginning of the First Century B. C. Their argument is directed

[33] Galen, I. 41, f., Kühn.
[34] *Analyt. Priora*, II. 23–24.
[35] *Philodem über Induktionsschlüsse*, . . . herausgegeben von Th. Gomperz, 1865; later editions and discussions by Bahnsch, Lyck, 1879, and Philippson, Berlin, 1881.

against a certain Dionysius, a Stoic otherwise unknown, who maintains the position of Aristotle as stated above. Zeno, the chief authority on whom Philodemus relied, was evidently a man of ability, who admired Carneades and was himself held in high esteem. He takes up one by one the Stoic arguments against the admissibility of inferences from experience, and, by drawing the distinctions which sharp analysis suggests, points out the conditions which must be met if the conclusions are to be valid. It is not necessary to discuss these in detail;[36] but it is of importance to note that in essentials these late Epicureans had hit upon the main principle of the modern procedure in experimental proof, which consists in controlling, confirming, or correcting one generalization by another, and in varying the conditions in order to test whether the attribute on which the conclusion rests is constant or not. The fact that this principle could be so clearly grasped and stated is sufficient evidence that it only formulates the process which men of science applied in practice.

Side by side with this development among the Epicureans ran another of no small significance. In the middle of the Third Century B. C. there grew up what is known as the Empiric School of medicine. Philinus, who is commonly called the founder of the school, which continued to exist at least until the Third Century of our era, was a pupil of the great Alexandrian physician Herophilus, and is said to have been greatly influenced by the scepticism of Pyrrho. It is a well-known fact that a large number of the later Sceptics were physicians, and that there was a strong bond of sympathy between the medical fraternity and the various schools which showed a sceptical tendency; but one may be pardoned for doubting whether the influences which

[36] Cf. Paul Natorp, *Forschungen zur Geschichte des Erkenntnisproblems im Altertum*, Berlin, 1884, p. 248.

brought them together really proceeded from the sceptical schools of philosophy. These undoubtedly furnished the weapons with which the Empirics assailed the doctrines of the Dogmatists; but that was in any case natural, unless the physicians were first of all philosophers. Herophilus took a peculiar position, which must have provoked serious thought on the part of his most gifted pupils, among whom we must reckon Philinus. He combined certain dogmatic assumptions, such as the Hippocratic doctrine of the four humors, with the most thorough research and the practice of dissection, to which he owed many epoch-making discoveries. At the same time he was a pronounced admirer of Hippocrates and devoted much time to the writing of commentaries on the Hippocratic treatises. Human nature being what it is, one can not well doubt that it was the new discoveries, due to his empirical methods, that most deeply impressed the pupil, especially since the Hippocratics themselves had set the example of close and painstaking observation. One can, therefore, conceive of the Empiric school as coming into being under the circumstances then obtaining, even if there had been no Pyrrho to cast doubt on the possibility of certain knowledge. The Empirics were in fact positivists rather than true sceptics; for they directed their doubts against the inferred causes and entities rather than the data of sense which fell under their observation. What they aimed in their writings to do was to discredit the hypotheses of the Dogmatists; but in reality they were building up an empirical science based on observation alone.

Their theory[37] recognized primary observation, or "autopsy," and investigation ("*history*"). There is no need of enlarging on the former; by "history" they meant the report

[37] Cf. Victor Brochard, "La Méthode Expérimentale chez les Anciens," *Revue Philosophique de la France et de l' Étranger*, XXIII, 39–49.

of competent observers, which, in case of doubt, should be confirmed by personal observation, if possible, several times repeated. Account must be taken not only of confirmatory, but also of unsuccessful experiments, and one must note whether the same remedy produces the same result invariably, generally, or only occasionally, and record the precise measure of success, otherwise the test is only partial and inconclusive. When one has often performed the experiment successfully, one arrives at a "theory;" and a collection of such "theories" constitutes an art. One must distinguish between properties of remedies and of diseases peculiar to them and such as are common to several. Symptoms are to be noted; symptoms being characters contrary to nature, some constant, others concomitant (but not regular). A disease is a group of symptoms, which arrive, continue, increase, diminish, and cease together. In addition to direct observation and the investigation of "history," one must, in cases not previously met with, have recourse to inference by analogy; but one is not to resort to a formula or general law. The procedure is precisely that which Mill calls inference from particular to particular. Menodotus,[38] an Empiric and Sceptic, probably of the First Century of our era, developed this old doctrine further, by emphasizing the importance of testing the results of the procedure based on observation and analogy by skilful experiment and by supplementary rationalization (*"epilogism"*). By the latter he seems to have meant the checking of one generalization by another, although he would not allow that the procedure yielded apodictic certainty. He also

[38] Cf. Albert Favier, *Un Médecin Grec du II*e *Siècle après J.-C., Précurseur de la Méthode Expérimentale Moderne: Ménodote de Nicomédie,* Paris, 1906; Albert Goedeckemeyer, *Die Geschichte des Griechischen Skeptizismus,* 1905, pp. 257 ff.; Paul Natorp, *Forschungen zur Geschichte des Erkenntnisproblems im Altertum,* 1884, pp. 156 ff.

formulated in some detail the canons of evidence to be applied to the testimony of others in cases where the individual could not himself observe the phenomena in question.

Such in outline is the position to which empirical science had attained at the close of ancient times.   Mill has said:[39] "The induction of the ancients has been well described by Bacon under the name of 'Inductio per enumerationem simplicem, ubi non reperitur instantia contradictoria.'   It consists in ascribing the character of general truths to all propositions that are true in any instance that we happen to know of.   This is the kind of induction which is natural to the mind when unaccustomed to scientific methods.   The tendency, which some call an instinct, and which others account for by association, to infer the future from the past, the unknown from the known, is simply the habit of expecting that what has been found true one or several times, and never yet found false, will be found true again.   Whether the instances are few or many, conclusive or inconclusive, does not much affect the matter; these are considerations which occur only on reflection; the unprompted tendency of the mind is to generalize its experience, provided this points all in one direction; provided no other experience of a conflicting character comes unsought.   The notion of seeking it, of experimenting for it, of *interrogating* nature (to use Bacon's expression) is of much later growth.   The observation of nature, by uncultivated intellects, is purely passive; they accept the facts which present themselves, without taking the trouble of searching for more; it is a superior mind only which asks itself what facts are needed to enable it to come to a safe conclusion, and then look out for these."   It is not clear how much of this statement was intended to apply to the ancient Greeks; but one has reason, from remarks made in many "standard" works, to

[39] *Logic*, Bk. III, Ch. IV, §2.

conclude that it has been accepted as the final judgment of the great logician on the attempts of the Greeks to lay the foundations of Science. That it falls far short of doing them justice is so plain that there is no need of further comment.

III. CLASSIFICATION

"The first steps of most of the sciences are purely classificatory."[1] This is not surprising, since classification is the first form taken by understanding. Even the most directly sensory images are grouped and, when language is employed, are designated by nouns. How strong the tendency is is shown by the fact that even words intended as proper names, like "Papa," "Mamma," become common nouns to the child, and may be used generally of any member of a class. Every abstraction or generalization is a step toward classifications, and every judgment implies it.

Classification is simply the ordering of experience, and, like every order, is due to the movement of attention directed by an interest. Since interest is primarily utilitarian, classification in general serves a practical purpose; when science matures and draws away from clamoring wants that brought it into existence, it sets up ends of its own, which may be far removed from those of our daily life; but the classifications it then forms are no less the expression of its ruling interests. One of the difficulties which beset all thinking is that classifications made for different purposes and with different interests exist side by side because they are a part of our intellectual stock in trade and are firmly established in language; and once they have become common property it is quite impossible to get rid of them.

Classification underlies definition and generalization and, since these constitute the frame of knowledge, science necessarily builds upon them. It would serve no useful

[1] James, *The Principles of Psychology*, II. 647.

purpose to discuss classification in general at greater length;
we shall therefore take up a few particular cases which have
a special interest for the student of science.

According to the tradition, which goes back to Aristotle,
the first scientists and philosophers of Greece began by
searching for the "elements" and "principles" of the world.
These would, in effect, be the ultimate categories under
which all things would fall. The terminology used by Aristotle is open to serious question from the historical point
of view; but we are not called upon to discuss that aspect
of the matter at present. We are rather concerned with
the question how the pioneers came to assume the precise
"elements" which they posited. One can hardly speak of a
single substance as an "element," when such thinkers as
Thales and Anaximenes derived all things from water or
air; the term first becomes intelligible when it is applied to
a number of substances which in various combinations constitute the concrete objects with which we are familiar.
The forms of matter which were usually called "elements"
by the Greeks are, as every one knows, earth, water, air and
fire. Unfortunately we do not know by whom this classification was originated. The earliest of the Greek philosophers to whom we can certainly trace the use of it is Empedocles; but everything points to the conclusion that it was of
popular origin and simply taken over by the philosopher-
poet. Taking a general view of the world, it was not un-
natural for one to divide it into land, water, air and the fire
which showed itself in the sun and stars. If, as seems certain, this was the origin of the so-called "elements," it is
evident that the classification is based on a very rough
division of the world into masses, which were not only
separable but on the whole actually separated and arranged
in scale according to density, beginning with earth as the
densest and lowest, and ending with fire, the rarest and

highest.   There is good reason[2] to believe that quite early
this arrangement was explained by the tendency of things,
observable in whirlpools and whirlwinds, to be deposited in
the order of density (or size), the coarsest and densest at the
center, the finest and most subtile at the periphery of the
circle.   It is obvious, however, that these four "roots" (to
use the name which Empedocles employed) were not found
by any process of scientific analysis; they were really nothing
more than aggregates, which might serve the purposes of a
grouping.   Nevertheless confusion and difficulty did result
from this attempt at simplification.   The earliest Greek
thinkers might say that all things came from one—water,
say, or air; but they had no more distinct idea of what constitutes a thing *one* than of what constitutes it an *element*.
When the Eleatics insisted that in order to be *one* a thing
must be *homogeneous*, they introduced a new mark, which
raised a whole crop of difficult questions.

First of all, how could there be any change?   If all were
one (and homogeneous), how could it differentiate itself?
One part being, by hypothesis, like every other, why should
any introduce a change in the whole?   If the all were many,
the "many" must be composed of units which were in themselves homogeneous but differing from one another; in
which case, having nothing in common, the different units
of one "root" could not act upon those of another sort, even
supposing that any action could arise in themselves.   It is
not necessary to go into the solutions which different
thinkers offered for these and similar difficulties—suffice it
to say that in the effort to meet them ideas were framed
which answered fairly well to the modern conception of an
element.   In the intellectual struggle, however, the four so-
called "elements" lost their truly elemental character; for
they were broken up and gave way to other classificatory

[2] Aristotle, *De Caelo*, II. 13, 295 a 10 ff.

groups.   Where they continued to be regarded, it was generally only as *maxima mundi membra* that they functioned, or
(as with Aristotle and the Stoics) they were supposed to pass
one into the other.   It is one of the most striking examples,
however, of the persistence of popular notions and terminology that even scientists continued to speak of the four elements, although the fact was well known among them that
earth contained many distinct kinds of earths, and that
water, as it occurs in nature, is far from uniform.   Every
potter knew the former fact, and the physicians never tire
of insisting on the latter.   Anaximenes knew different
states of air, and Aristotle distinguished between the fire
which we use and the fire of the stars.   It seems to be characteristic of the mind that it insists on retaining its distinctions and classifications even when other and better ones
have been found.

The division of the material world into the so-called elements was, psychologically considered, an effort at simplification.   Whether scientific or pre-scientific in origin, it
was based on observation; but the observation was not
purely passive.   Though the phenomena in question might
be seen in the action of whirlwinds and eddies, it is clear
that there was a certain amount of rough experimentation
carried on, such as stirring gruel in a bowl or muddy water
in a pail.   It was known that the gruel separates if it is not
stirred,[3] and stirring, then as now, was generally by making
the liquid move round and round, producing a vortex.[4]
Then came the further generalization which included the
cosmos in the process, for it was assumed that the separation of the "elements" was due to the revolution of the
heavens.   A further conclusion followed, to wit, that the
cosmos was a series of concentric circles, the earth at the

[3] Heraclitus, fr. 125, Diels.
[4] For example, Hippocrates, *De Morbis*, IV. 55 (VII. 600 L.).

center, surrounded by the river Oceanus, which encompasses it, followed by the circumambient air and, outermost of all, by the "flaming battlements of the world," the ring of fire composed by the celestial luminaries. The "battlements" might be numerous, but the picture still held. Such was the conception entertained, for example, by Anaximander. It was a bold generalization, setting aside the popular notion of the heavens as an inverted bowl, which moreover had the support of the Homeric epic; but it was commended by the fact that the phenomena in question may be observed only on a plane. There was nothing in them that suggested a series of spheres.[5] We may see in this early theory an anticipation of many modern theories, beginning with the nebular hypothesis; but what especially interests us is the fact that it held the field until purely mathematical theory, harking back to popular notions, once more conceived the heavens as a sphere and deduced the sphericity of the earth.

Meanwhile the first scientific theory had at least so much in common with popular notions, that it regarded the earth as a disk.

The difficulty presented by the observable fact that the heavenly bodies move *over* the earth was met by the theory of the "dip" of the earth to southward; for the early scientists contended that originally the sun had moved round the edge of the earth-plane. The outer circles of air and fire were clearly regarded as broad belts surrounding the earth at a sharp angle caused by the "dip," which moreover accounted for the seasonal changes, since the earth was thought to dip more to southward in summer than in winter. Into these matters it is not necessary to go at greater length; suffice it to say that, when the spherical

[5] The loose use of "sphere" for "circle" in our sources has led to misunderstanding of several early thinkers.

heavens and earth became established, the dip gave place first to the obliquity of the ecliptic, and then, after Heraclides of Pontus, to the inclination of the earth's axis to the ecliptic.

One of the greatest of Greek achievements in classification was the creation of scientific geography;[6] for, of course, descriptive geography is nothing more than a classification of places in terms of spatial relations, comparable to an analytical chart. We can not rehearse the history of this science, though it is most interesting and well illustrates the scientific procedure; a few points, however, may serve to indicate the nature of the problems and the way in which they were solved. Topographic charts are known to have been made in Egypt and Babylonia at a comparatively early date; but comprehensive charts of the earth were apparently first made by the Greeks. No doubt, rude sketches of trade-routes and rutters were attempted by merchants and sea-farers even before we find hints of them in the *Odyssey*, where the general directions and distances of lands in the eastern Mediterranean are clearly indicated. We also read of sailors guiding their course by the stars, though in general they skirted the shore-line. Such was the situation when Anaximander of Miletus, the swarming hive of colonizers, before the middle of the Sixth Century B.C. attempted the first general map of the then known earth. Of its merits we have no correct means of judging; but we are informed that Hecataeus, also of Miletus, in the next generation improved his map wonderfully. This might be true, however, and yet not mean that Anaximander's was not good; for Hecataeus possessed a knowledge of even the western Mediterranean which, for the time, is truly amazing.

[6] See my study, "Anaximander's Book, the Earliest Known Geographical Treatise," *Proceedings of the American Academy of Arts and Sciences*, LVI (1921), 239–288.

The information necessary for the construction of the charts —they were essentially charts, dealing chiefly with the coast lines, which were best known to the sea-faring Greeks, but taking in the hinterlands as well—was obtained partly by personal inspection, partly by inquiry and report. Report was of course not altogether trustworthy, being in good part fabulous. Hakluyt's *Voyages* and the early maps of the New World will give the modern reader a notion of what the Greek pioneers had at their disposal.

Unfortunately all ancient maps have been lost; but there are means at hand for reconstructing them from the later geographical literature. Any classical atlas will be found to give such reconstructed charts, though some of them leave much to be desired. We can not go into details, but must content ourselves with merely indicating the methods used by the ancient geographers. It must be borne in mind that the instruments now regarded as indispensable had not then been invented—there were no magnetic compasses nor chronometers. At first the sextant also was unknown, though sun-dials were in use. It is important to note that the latter instrument was serviceable in reckoning time rather than in determining position; and the time was simply the time of day or the season, not serving at all for dead-reckoning. For the latter, the only means then existing was the water-clock or clepsydra, and that does not appear to have been used in reckoning longitudes before the time of Hipparchus.[6a] By combining the readings of the clepsydra and the sun-dial, a rough approximation of longitude could be obtained; and evidently Hipparchus achieved all that was possible with such means. The sun-dial was

[6a] On land, of course, distances were measured, the longer ones by day's journeys. Probably from the time of Eratosthenes onward certain data were known regarding the difference in time of eclipses; but they were too few and inaccurate to be of much service to the geographer.

used for determining latitude, but at first it was misleading rather than helpful. Very primitive peoples have noted the changes in the position of the sun and have constructed dials, sometimes of vast scale, marking the risings and settings at equinoxes and solstices. Quite apart from the fact that these points are most easily marked on the horizon, the angles also are greatest there; hence it was at first the position of the sun on the horizon that was noted. On the other hand, though the length of the shadow cast at noon by the gnomon was indicated on the dial, its significance for latitude was not at once appreciated. We have no evidence of attempts to determine latitude by observations of the meridian height of the sun before the Fourth Century. This is not surprising, although the fact seems to have been quite overlooked, because the true reason for the curve plotted on the dial by the sun between rising and setting could not be understood before the earth was known to be a globe. Before that discovery the position of the sun on the horizon was naturally taken to be most significant; and, as has already been said, it led to a natural but serious error. The observant Greeks of course noted the fact that the points at which the sun rises and sets at the solstices are equidistant from the equinoctial points, which lie due east and west; but they did not know that that must be true of any point on the surface of the globe. Consequently they supposed that it was true only of their own region and constructed their maps accordingly, giving the cardinal points of the dial a geographical reference.

One can, therefore, understand one feature of the early maps. In locating the coordinates, the equatorial ran through Ionia west to the Pillars of Hercules, or the Straits of Gibraltar. This is true, apparently, of all the later maps, even down to the time of Columbus, though of course the true equator had long since been duly located—such is the force of tradition.   The origin of coordinates appears in the early period to have been Delphi, "the navel of the earth;" but by the time of Herodotus, in the Fifth Century, one of the chief meridians, if not the chief, was that supposed to run from the Pelusiac mouth of the Nile through the Cilician Gates and Sinope to the mouth of the Ister (Danube).   From Herodotus we learn that the map which he used located the summer solstice (the tropic of Cancer) on a line from the mouth of the Ister to the Pyrenees, and the winter solstice (the tropic of Capricorn) at a corresponding distance southward, following the line of the supposed bend of the Nile below the First Cataract westward to the Atlas Mountains.   One can well imagine the shock caused by the discovery, known to Eratosthenes, in the Third Century, that the tropic of Cancer actually was situated almost exactly where the tropic of Capricorn had first been located. Such a reversal was possible only because earlier geographers had not perceived the significance of the meridian height of the sun.   Another consequence of the failure to use the sun-dial to the best advantage was the distortion of the maps in certain regions.   There being available no means of accurately determining latitude, certain "climes" were fixed, which were actually in part defined by climatic conditions.   It is no wonder, therefore, that in regions little accessible, a considerable dislocation resulted.

If one traces on a modern map the supposed meridian running from the Pelusiac mouth of the Nile to the mouth of the Ister one sees that it follows a devious course; but, considering the means available for determining longitude, one has good reason to admire the degree of accuracy attained.   Besides the absence of compass and chronometer there was the want of a coast-survey.   The reported length of the coast-line of Egypt on the Mediterranean[7] is not rec-

[7] Herodotus, II. 6–7.

oncilable with the facts; elsewhere distances were generally
given in terms of a day's march inland or a day's run of a
ship at sea. Directions were roughly indicated, generally
by the cardinal points or at sea by the direction of a following wind, because tacking was unknown.[7a] The wind-rose
rather than the finely divided compass-card set the standard;
and, though Aristotle knew that the dividing line of the
wind-rose practically ran from northwest to southeast,[8] no
account of this fact seems to have been taken in practice.
Where close personal observation was out of the question
for the geographer, he had to rely on ships' logs, which
would give the number of day's runs, from which a rough
estimate could be deduced of distances and the course followed, which would be approximated to the cardinal winds.
That nevertheless a reasonable accuracy was secured testi-
fies to the care with which the pioneers must have checked
such evidence as they could obtain.

In the Fourth and Third Centuries B.C., a great improvement came about. Not only were lands hitherto little
known opened to Greek men of science by the extended
conquests of Alexander and the ordered empires of his successors, who everywhere sought to enlist the most intelligent Greeks in their service, but the scientific advances of
the Fifth and Fourth Centuries had made clear the nature
of the problems and taught men how to make full use of
the instruments they had long possessed. Sun-dials now
began to be used in determining latitude and, with the
water-clock, provided the data for reckoning longitude. It
is plain that in the parts best known to the Greeks, the de-

[7a] Homer, *Od.*, X, 32, and *Hymn to Apollo*, 405, sometimes cited as evidence of tacking, do not prove knowledge of the practice, which was certainly familiar to Lucian, in the Second Century of our era.

[8] The winds were grouped, the north and west, and the east and south,
winds being regarded as paired.

terminations made by Hipparchus about the middle of the Second Century must have been very accurate; for Marinus and Ptolemy could find little fault with his maps, except in remoter regions, where he generally followed the old Ionian charts.   How much use Hipparchus may have made of trigonometry (which he originated) in his geographical treatise we have no means of knowing.   Of the subsequent development of descriptive geography it is not necessary to speak.   All students of history know the part which it played in the discovery of the New World.

We are told that Anaximander, who first essayed a map of the known earth, matched it with a celestial "globe." There is the best of reasons for believing that this is a misnomer; if, as is not unlikely, he did chart the stars, it is all but certain that he did so on a plane surface.   We should, no doubt, be greatly interested in the list of constellations which he included and the positions he assigned to them. He is credited with the discovery of the obliquity of the ecliptic, which may or may not be true.   Certain constellations had long been distinguished and their risings and settings had served to mark the seasons in the farmers' and sailors' almanacs.   Though the position of the sun among the constellations could be noted only at the risings and settings of the zodiacal signs and by inference, the retrograde course of the main planets with reference to the apparent motions of the fixed stars and the limits of their aberrations could be directly observed, and were early noted.   How accurately the star-maps were drawn we can not say, because we do not know anything in detail about the early instruments.   We know, of course, that they had something roughly corresponding to our transit or theodolite in the Fifth Century, and we hear of observatories, which were probably merely observation-posts, even earlier. Sighting-instruments were certainly in use by engineers as

early as the Sixth Century; but we know nothing in detail
about the manner in which they were employed in astronomy. The Babylonians[9] from the Seventh Century onward had developed a very accurate system of notation by
which the location of a star could be described and their
star-charts must have been correspondingly exact; but we
have no evidence of a knowledge of Babylonian methods on
the part of the Greeks before the time of Alexander the
Great. Even earlier certain astronomical data, even
rather complicated ones, were known to astronomers, such
as Meton; and later the exact observations and calcula-
tions of Cidenas (Kidinnu) were adopted by the astrologer
Vettius Valens. Some of the determinations formerly
attributed to Hipparchus have recently been discovered to
have been borrowed by him from Cidenas; but his elaborate
chart of 850 stars and the discovery of the precession of the
equinoxes[10] were probably his own work, however much he
may have learned from the Hellenized Chaldaeans, like
Cidenas. Eudoxus had, probably, first constructed a celestial globe, and Ptolemy added many to the number of
stars charted by Hipparchus. This required much accurate observation, but the precise character of the instruments employed can only be inferred. However, it is
fairly certain that the mediaeval astrolabe closely followed
Greek models, which must have differed little from the
Babylonian. In descriptive astronomy the Greeks owed
more to the Orient—particularly to Babylon—than in any
other branch of science; but even here they were not merely
receptive.

[9] Cf. Franz Cumont, "Babylon und die Griechische Astronomie," *Neue
Jahrbücher für das Klassische Altertum* etc., XXVII. (1911), 1–10.

[10] This discovery has lately been claimed for the Babylonians; but
Schnabel's inference, which would assign the discovery to Kidinnu, known
to the Greeks as Cidenas, is not accepted by Kugler, II. 592 ff.

In cosmography, astronomy, and geography the Greeks made the effort to define their position in space; no less important was their attempt to determine their temporal relations.   We might here take up their gradual development of a scientific calendar, but in this regard they could achieve nothing that materially improved what had already been done by Babylonians and Egyptians.   We shall, therefore, confine our attention to another aspect of chronology. When in the Sixth Century the Greeks began to reflect seriously, they realized that they were a young people[11] surrounded by nations whose origins were lost in antiquity. They had, you might almost say, no history; what passed for such was legend, and that did not carry far.   It might not unjustly be described as *Tales of a Grandfather*, for in general it took the form of family-trees, to which at important points traditional details were added of events of greater significance.   Such genealogies naturally were treasured in the families which either still ruled or had lately ruled their respective cities.   Their eminence was probably in most instances due to the part their forebears had played in stirring events which lived on in legend and served to articulate it.   Great houses frequently intermarry, and consequently family-trees interlaced, establishing connections which in the aggregate rendered possible a chronology based on an average generation.   By giving the generation a numerical value, a consistent and approximately accurate history could be reconstructed, working backward from the present.   That precise procedure was followed in the Sixth and Fifth Centuries, and probably earlier.

But the result was not such as could satisfy a Greek.   In the first place, the framework was Greek and the system was essentially self-contained.   It sufficed to fix within certain limits the era of the Fall of Troy, but the Greek's

[11] Cf. Herodotus, II. 142 f.; Plato, *Timaeus*, 22 b f.

awakened sense of the immensity of the world in time, as well as in space,[11a] could not content itself with so limited a scope.   There was no general chronological scheme anywhere available.   The era of Nabonassar (747 B.C.) might serve for Babylonian chronicles, and, if it were actually recognized, the Sothic Period might be useful in determining dates in Egyptian history *within a particular period* of 1461 years, but in general events were dated by regnal years of a sovereign, the beginning of each reign marking a new era. Greek methods of reckoning by Olympiads, beginning with 776 B.C., by priestesses of Hera at Argos, by Athenian Archons, were for practical purposes as good; and the era of the Fall of Troy might compare favorably with the best. Indeed, it possesses for us the added value of being a date inferred by calculation instead of one objectively given.

In the Sixth Century there was made an attempt to establish an absolute chronology for universal history, which for scope and ingenuity equals the most daring enterprises of the human mind.   Unfortunately we are in no position to say how many cooperated in the undertaking, and the details at our disposal are few; but they suffice to reconstruct the scheme in outline.   To all appearance the originator, at least, of the scheme was Hecataeus of Miletus, the same man who made so notable a contribution to geography.   He wrote a great work, or possibly two works, the first part being in later times entitled *Genealogies*; the second, *Descriptive Geography*.   As the latter embraced the whole *orbis terrarum* then known, one might reasonably conjecture that the former was not confined to Greek family-trees; indeed, it is fairly certain that Herodotus derived the greater part of his lists of Egyptian kings from the *Genealogies* of Hecataeus, and that the notorious break[12] in

[11a] Cf. Plato, *Theaetetus*, 174 b ff.
[12] After II. 141.   I hope ere long to treat more fully of this matter.

the account of Egypt given by the "Father of History" is due to no other cause than an awkward attempt to piece together data which Hecataeus had for reasons of his own distributed among the parts of his comprehensive account. In that work Egypt is known to have occupied the center of interest and to have received a treatment quite out of the usual proportions.   Owing to want of information in detail it is not always possible to say just how much of the scheme which we are enabled to trace is due to Hecataeus himself and how much may have been added by others.

Hecataeus may be assumed to have gone to Egypt in the train of Cambyses in 525 B. C., or not much later; for he was clearly one of the "elder statesmen" at the time of the Ionian Revolt in 499.   In Herodotus we have a singularly interesting datum: we are told[13] that "Hecataeus the historian was once at Thebes, where he made for himself a genealogy which connected him by lineage with a god in the sixteenth generation.   But the priests did for him what they did for me (who had not traced my own lineage). They brought me into the great inner court of the temple and showed me there wooden figures which they counted up to the number they had already given, for every high priest sets there in his life-time a statue of himself; counting and pointing to these, the priests showed me that each inherited from his father; they went through the whole tale of figures, back to the earliest from that of him who had lateliest died.   Thus when Hecataeus had traced his descent and claimed that his sixteenth forefather was a god, the priests too traced a line of descent according to the method of their counting; for they would not be persuaded by him that a man could be descended from a god; they traced descent through the whole line of three hundred and forty-five figures, not connecting it with any ancestral god

[13] Herodotus, II. 143, tr. Godley.

or hero, but declaring each figure a 'Piromis' the son of a 'Piromis,' that is, in the Greek language, one who is in all respects a good man." This is a remarkable statement, for we see that the generations of priests are precisely parallel to those of the Egyptian kings.[14] We thus have 345 generations of human history in Egypt, and, counting three generations to a century, 11,500 years.

The whole scheme certainly derives from Hecataeus, to whom we must also assign the attempt, which follows immediately in Herodotus, to establish a connection between Greek and Egyptian gods and heroes. This was no easy task, because Greek mythology assumed far more recent dates for the doings of its personages on earth; duplication had accordingly to be assumed for Dionysus and Heracles, though the precise chronology to be adopted for the deified heroes bearing those names in Greek myth and legend might be open to dispute. The first desideratum, however, of an absolute chronology was satisfied; for the millenia of recorded history in Egypt afforded ample room, it seemed, to accomodate all the events, such as the periodic destruction of civilization by flood or fire, which Greek legend reported. It was thought notable that in all these catastrophes Egypt had remained untouched, thus presenting an unbroken record which might account for the high civilization one saw evidenced on all hands in the marvelous monuments of the past. Nevertheless an absolute timescale would be of no use unless points of contact could be established between Egyptian kings, who appeared in the lists and could accordingly be dated, and personages of Greek story. Where could one find such synchronisms? Well, there was the Homeric story of Menelaus and Helen being driven to Egypt on their return from Troy. If one could assign this event a place on the absolute scale, one

[14] Compare Herodotus, II. 142.

had at any rate a definite connection not only for the all-important epoch of the Fall of Troy, but also a point from which one might reckon forward and backward by generations and determine other dates. Yes, Helen had met Proteus, and Proteus was an Egyptian king,[15] a man of Memphis, who succeeded Pheros. Thonis, of whom the *Odyssey* spoke, was only a local lord ruling in the Delta. It is not necessary to go into more detail here, because it was known that contact with Egypt had been constant in recent times from Psammetichus onward.

Thus was created a chronological scheme which connected the history of the Greeks with what they thought the most ancient people of the earth; but that was not yet universal history. How were the other great nations to be brought into relation to the general scheme? Perseus was a Greek hero, but he was obviously the eponymus of the Persians. Though he could not be directly connected with Egypt—he had no known father—he was so connected indirectly, because Acrisius, father to his bride Danaë, was of Egyptian descent.[16] Babylon was likewise drawn into the complex, in that Belus (Bel), its founder, was made an Egyptian. Further help was obtained by availing oneself of the deified heroes Dionysus and Heracles, who wandered over many lands and sired numerous royal families. The "Egyptian Heracles" in particular played an important rôle in binding together the ends of the earth by the royal

---

[15] Herodotus, II. 112. The Greeks generally dated the Argonautic Expedition two generations before the Trojan Era, and Herodotus II. 104 represents the Colchians as a remnant of the army of Sesostris, the second predecessor of Proteus. That this reconstruction of history was not due to Herodotus, but was unintelligently borrowed from another, is shown by the facts that Proteus was in the eleventh generation, say 400 years, before Amasis, which Herodotus can not possibly have taken for the Trojan Epoch. He dated Homer and Hesiod about that time (II. 53).

[16] *Ibid.*, VI. 53.

houses and connecting them with his native land. One
might almost fancy one was reading Mr. Perry's story of
the Sons of the Sun.

No doubt, all this is extremely fantastic. Not only are
the data mythical or legendary, where we require really
historical facts, but they are dealt with in very arbitrary
fashion. A modern historian, one should think, would be
laughed out of court, were he to produce a similar reconstruction of history, though one need not go far in search
of parallels. But reflect. Historical criticism was not yet
born; and myths and legends were naturally regarded as
data which one might well use, if one only rationalized
them. Here, too, one need not go back to the Sixth
Century B.C. for illustrations of the same procedure.
Moreover, such were the only data available. Archaeologi-
cal research and the decipherment of ancient documents
unknown—and, had they been known, unintelligible—to
the Greeks have made it possible in our time to lay more
solid foundations; but, if one regard only the mental processes involved in creating the framework of universal history and in establishing the synchronisms necessary to
articulate its various members, one recognizes that they
are the same by which we are today extending the domain
of history.

We can not here enlarge upon Greek historiography; but
it is important to note that the impulse given in the Sixth
Century continued to make itself felt throughout ancient
times. Thucydides, by choosing a limited period and re-
cording contemporary events, set an example which some
notable historians followed; but the ideal of universal history seems on the whole to have prevailed. Herodotus
belongs to this line of tradition. He did not, like Hecataeus,
take Egypt for the background of his story, which was
meant to deal principally with the Persian Wars. Whether

he was influenced by Charon and Dionysius we have no means of knowing; but, if we were to be guided by the main thread of his account, we might entitle his history, like that of these writers, *Persica*. Egypt and other lands claim his attention as they are successively reached in the expansion of the Persian Empire. The most recent history naturally received the fullest treatment, a practice which formed the rule in later times. Another important characteristic of universal history, which is due to the pioneers, is that history and descriptive geography, including ethnography, were habitually combined; and by the more philosophically minded historians and geographers their proper theme was prefaced by a cosmographic sketch. Thus, in conception at least, the things of here and now were to be presented in perspective; the historian, like the true philosopher, became a spectator, as it were, of all time and all existence. Historical science has made great advances in the last generation or two, but the improvements have been due chiefly to the refinement of technique, made possible by the vast extension of knowledge and by the development of ancillary sciences. The pioneers defined the ideal and in a limited way exemplified the methods which are in use to this day.

Classification in the more ordinary form might be illustrated by many examples drawn from medical literature. The four humors and the corresponding temperaments at once suggest themselves; and of course diseases were grouped and distinguished from time immemorial. Folk-medicine is apt to draw few and rough distinctions and treat all cases falling into its classes by applying the favorite nostrums. That was the stage at which medical practice had arrived at Rome when Greek physicians first came and were greeted with the threats and denunciations of the elder Cato. In Greece the Coan and Cnidian schools seem to have differed, among other things, in this, that the latter carried its classification very far, subdividing species into minutely differentiated varieties, whereas the former recognized more general groups while insisting that the treatment must be governed by regard for the observed peculiarities of the individual case. How great the distinction would prove to be in practice it is not easy to tell.

The classification of foods and medicaments depended, of course, primarily on their physiological action,[17] which was carefully observed. Details would be of interest only to specialists. There is, however, a table of foods given in the *Second Book on Diet*[18] which may well claim our special attention. They are divided into (I) foods derived from plants, including barley and barley-gruel; wheat and various kinds of wheat-bread; spelt and oats; barley-groats, fresh and old; pulse, including various kinds of beans, peas, chick-peas, etc.; (II) food derived from animals, including quadrupeds; birds; fish, crustaceans and testaceans; eggs; cheese; (III) beverages, including water, wine, must and vinegar; honey; (IV) herbs and fruits, including, under the former head, almost all of the vegetables still in use, and, under the latter, not only orchard-fruits but also such things as cucumbers, etc. The classification here given makes no claim to completeness or scientific accuracy, but is conceived in a purely practical spirit. It serves merely to group related things together for the purpose of enumerating their uses. This is made the more evident by the position of IV in the scheme, for it naturally would come under I. The reason for its postponement is presumably to be sought in its relative unimportance from the point of view of the practising physician. Nevertheless, it is interesting

[17] Hippocrates, *De Victu*, II. 39 (VI. 534 f., L.).

[18] *Ibid.*, II. 39–56. For this, see I. Klüger, "Die Lebensmittellehre der Griechischen Ärzte," *Primitiae Czernovicienses*, II. (1911), 1–53.

to observe the generally clear order which the author establishes for the systematic discussion of the properties of the foods most commonly prescribed.   It would be unfair to demand of him as logical an analytical scheme as one finds, for example, in the biological treatises of Aristotle.

Even in logic, however, classification is not necessarily strictly analytical; for Aristotle's Categories are evidently based on an empirical survey of the different kinds of predication.   His four-fold causation, however, rests upon a nice analysis of the process of realizing a purpose.   More elaborate examples of classification based upon extended research and analysis may be found in Plato's and Aristotle's theoretical discussions of the State.   The results may be assumed to be more or less familiar to all serious students, and need not be stated.

Aristotle was well aware that induction depends upon classification,[19] if indeed they are not at bottom the same. Definition likewise is only a special form of classification. Plato in the *Sophist* and the *Statesman* exhibited a method of arriving at a definition by systematic division, which was followed with minor changes by the Schoolmen.   In the *Republic*[20] he employed the method of elimination which was familiar in mathematics; and continually in his logical and metaphysical dialogues he emphasized the need of determining the mutual relations of ideas, a task which, had he accomplished it, would have presented a complete conspectus of the phenomenal and intellectual worlds.   In general, however, classification among the Greeks bears upon its face the evidence of its origin in practical life, its character clearly being dictated less by theoretical considerations than by the matter for the moment taken in hand.

[19] *Topica*, VIII. 2.
[20] 427 e f.

IV. ANALOGY

Psychologists are agreed that the ability to detect similarities is one of the most valuable qualifications for the extension of knowledge and the integration of the ideal world which is the creation of thought. The recognition of analogies is, therefore, a matter of consequence in the development of science. The genius has been defined as one who is extraordinarily apt at discovering resemblances which escape the notice of most men; and it is to the genius that, first and last, progress in whatever field is due.

Noting an analogy, however, implies the recognition of difference as well as agreement, although the latter may for the moment appear to receive more emphasis. In this sense the operation prepares the way for classification. It were little to our present purpose to discuss analogy in all its bearings, but there are two aspects of it that deserve special attention. One of these relates to the framing of hypotheses, the other to the confirmation of them. The mental processes of inference are anything but simple, and perhaps the only general statement one may safely make regarding them is that they do not in the first instance assume a syllogistic form. More probably thinking passes directly from particular to particular on the basis of an observed or assumed similarity: the inference is implicit in the analogy that suggests itself to the mind. Every scholar must recall many experiences of the shock which comes when, in considering or discussing an obscure subject, a parallel suddenly presents itself to the mind and at once assumes the form of an hypothesis; the inference is in fact already drawn by the mere recognition of the analogy. Or, again, a theory may have been proposed, by oneself or by another, and the noting of a parallel case may promptly confirm or invalidate the inference suggested, because, as was remarked above, the analogy is by its nature marked as much by difference as by agreement.   If in the reflective survey following the comparison, the similarities outweigh the differences, the first hypothesis is apt to be confirmed; if not, it will be modified or rejected.   In the absence of definite data we reason from probabilities, and in many matters the only probabilities available are such as are suggested by analogy.

An instance that is very much to the point is the sort of explanation offered by psychologists on the basis of the theory of psychophysical parallelism.   To be exact, the precise facts are known regarding neither the mental nor the physiological processes, but they are known or assumed to be parallel; and, since the latter are physical, they are explained on the analogy of mechanical systems, such as a telephone-exchange with its complicated devices for giving a specific direction in response to an outside call.   The elaborated analogy serves to explain immediately the physiological and mediately the psychological process.   Although the conscientious scientist will recognize and acknowledge the inconclusiveness of such reasoning, he may well maintain that it is both justifiable and useful.

Greek medicine owed its successes largely to the close observation of the patient.   The symptoms to which it attended were such as present themselves to the eye or may be ascertained by questioning.   It benefited by the exposure of the body in gymnastic exercises and relied in great part on dietetic treatment, endeavoring to suit kind and quantity of nutriment to the needs of the individual. Surgery developed to a high stage of efficiency in certain directions, such as the reduction of dislocations and the treatment of wounds; but dissection of the human cadaver and, especially, vivisection were little practiced before the time of the Ptolemies. When one dissected animals for scientific purposes, one did so on the tacit or express assumption of a close analogy between the human and the brute anatomy, and various errors naturally resulted therefrom. Similarly false inferences were drawn from observations made on the dead body, such as the conclusion that the arteries carried not blood but gas (pneuma). In a word, the fatal weakness of Greek medicine was the result of an inadequate development of physiology. Considering the restrictions imposed upon scientists by sentiments, which prove a deterrent even in our time, and by the want of microscopes and other means of investigation now at the service of the physiologist, we can not greatly wonder at the limited progress made by the pioneers in this direction. In principle, there is no difference between the situation of the ancient physician and the modern psychologist; such difference as one may note lies rather in the degree of complexity or simplicity of the phenomena with which they are concerned. In both cases the procedure followed is the same: where definite and assured knowledge fails, recourse is had to reasoning by analogy.

Galen, who for all his achievements was himself far from guiltless in this regard, animadverted upon this tendency among physicians: "One may find many such theories falsely set forth even by physicians who would not await the findings disclosed by dissection but stated conjectures founded on analogy as if they were based upon observation."[1] One must, however, in justice to the pioneers in any department of knowledge, bear in mind that precautions which now seem obviously necessary had to be enforced by experience before their importance could be recognized. Technique in every science is of slow growth, the cumulative

[1] *De Placitis Hippocratis et Platonis*, I. p. 165, ed. Müller.

creation of many workers extending and correcting the findings and methods of predecessors.   Today excavation by expert archaeologists is an art, almost a science; its operations are expected to conform to the most exacting rules of observation and record in detail.   It were foolish and highly unjust to demand of Layard and Schliemann the same accuracy and caution as regards inference as one rightly requires today.   In this field, as in physical experimentation, the importance of exact records was but slowly appreciated; indeed, it is more than doubtful whether the desirable degree of accuracy in this regard has even now been attained. The consequences, often deplorable, of the failure to take the required precautions may be traced in every science. Perhaps the fault lies in the very nature of our ordinary mental processes and in particular in the nature of attention, which is limited in direction and scope.   Even Faraday had to be told what to look for in an experiment.   When the immediate interest has been satisfied, and the particular phenomena for which we are looking have been noted, the task we have set ourselves is accomplished.   Whether we shall follow up the search either by taking note of other phenomena disclosed in the operation already completed or by further exploration is as nearly a matter of chance as anything one can think of; and, in case the quest is not continued, it is impossible to say what conclusions may be drawn from incomplete data.   One may cite as examples the descriptions of the vascular system given by Aristotle[2] after Syennesis of Cyprus, Diogenes of Apollonia, and Polybus the Hippocratic.[3]

It is worth while to quote the introductory words of Aristotle:[4] "Now, as the nature of the blood and the nature of

[2] *Historia Animal.*, III. 2.

[3] Cf. Littré, *Oeuvres Complètes d'Hippocrate*, IX. 164 f.

[4] *Historia Animal.*, III. 2, 511 b 10 ff., tr. Thompson.

the veins have all the appearance of being primitive, we must discuss their properties first of all, and all the more as some previous writers have treated them very unsatisfactorily. And the cause of the ignorance thus manifested is the extreme difficulty experienced in the way of observation. For in the dead bodies of animals the nature of the chief veins is undiscoverable, owing to the fact that they collapse at once when the blood leaves them; for the blood pours out of them in a stream, like liquid out of a vessel, since there is no blood separately situated by itself, except a little in the heart, but it is all lodged in the veins. In *living* animals it is impossible to inspect these parts, for by their very nature they are situated inside the body and out of sight. For this reason anatomists who have carried on their investigations on dead bodies in the dissecting-room have failed to discover the chief roots of the veins, while those who have narrowly inspected bodies of living men reduced to extreme attenuation have arrived at conclusions regarding the origin of the veins from manifestations visible externally." After quoting these earlier investigators he observes:[5] "The investigation of such a subject, as has been remarked, is one fraught with difficulties; but, if anyone be keenly interested in the matter, his best plan will be to allow his animals to starve to emaciation, then to strangle them on a sudden, and thereupon to prosecute his investigations." Aristotle then gives his own account of the vascular system, not distinguishing between veins and arteries, although the distinction was known to the Hippocratics.[6] We need not give details, which concern only the anatomist, but will content ourselves with quoting the comment of Professor Thompson:[7] "The Aristotelian account of the

[5] *Ibid.*, 513 a 12 ff.
[6] *De Carne*, 5 (VIII. 590 L.).
[7] Note on 513 a 35.

vascular system is remarkable for its wealth of detail, for its great accuracy in many particulars, and for its extreme obscurity in others. It is so far true to nature that it is clear evidence of minute inquiry, but here and there so remote from fact as to suggest that things once seen had been half-forgotten, or that superstition was in conflict with the results of observation." More probably the fault lay in insufficient records and the combination of observations made on different subjects. It is obvious that the dissection of animals supplied a good part of the data; some of the inaccuracies may be due to that fact.

The extent to which reasoning is based upon analogy is likely to escape our notice because we resort to it so constantly, and because it is often couched in simile or metaphor. Argument thereby becomes more concrete and clear, and therefore in general more convincing; that is one reason why the dialogues of Plato and the Socratic discourses in Xenophon are so effective. Arguing from analogy is, of course, not confined to political, ethical, and metaphysical inquiry; it frequently serves in lieu of, or in combination with, experimentation in confirmation of pronouncements concerning physical matters. This may be illustrated by a fragment of Archytas which is of sufficient interest to justify our quoting it for its own sake: "Mathematicians," he says,[8] "seem to me to have reasoned well, and it is not surprising that they have reached sound judgments about the nature of particular things; for, having reasoned well about nature as a whole, they were likely to see clearly in regard to the nature of things in detail. They have handed down to us clear knowledge of the velocities and the risings and settings of the stars, and about geometry, numbers, and spherics, and not least about music. For these sciences seem to be related, seeing that they are

[8] Fr. 1, Diels.

concerned with the related first principles of reality (to wit, number and magnitude).[9] First, then, they considered that sound can not arise unless certain things collide with one another. They said that an impact comes about when bodies in motion meet and collide. Bodies, therefore, moving in opposite directions, meeting (produce sound) by slowing down (*i.e.*, by checking each the other's motion of translation), while bodies moving in the same sense, but with unequal speed, overtaking one the other, produce sound when the slower is overtaken by the faster. Many of these sounds can not be taken in by our sense-organs, some by reason of the weakness of the collision, some because of its distance from us, some because of its excessive strength; for exceedingly great sounds can not enter our hearing, just as nothing enters small-necked vessels when one attempts to pour in a great quantity. Of the notes that reach our organs, those propagated by a violent collision come quickly and appear to be high; those that come slowly and weakly, deep. If one takes a rod and gives it a slow and weak motion, one will produce a deep note by the stroke; if quick and sharp, a high one. Not only by this may we know it, but in speaking or singing if we wish to sound a loud and sharp note, by uttering it with a strong breath (we shall succeed; if we wish to produce a deep note, we shall use a weak breath).[9] Furthermore, one may observe the following also, as in the case of missiles; those violently hurled, carry far; those weakly thrown, but a short distance. For the air yields more to things violently borne, but less to those weakly impelled. The same thing will happen in the case of sounds: a note carried by a violent breath will be strong and high; one by a weak breath, weak and deep. By this most certain token also we may see that if the same person utters a loud note we should hear him far; if a weak one,

[9] Words in parentheses supplied by Diels.

not even hard by.   Moreover, it is the same in the case of
flutes.   The breath from the mouth, if it enters into the
holes nearest the lips, by reason of its strength gives off a
higher note; if into the farther ones, a deeper.   Consequently it is evident that a rapid motion produces a high,
a slow one a deep note.   The same is true of the bull-roarers
whirled in initiation-ceremonies: moved slowly, they give
off a deep note; violently, a high one.   In the case of a reed,
also, if one closes the lower end and blows into it, it will
give a deep note; if one stops it at the half or any other
point it will sound sharp and high, for the breath passes
weakly through the long, strongly through the shorter distance.''

In such a case as this one can not tell—indeed, it makes
no difference—whether the principle suggested (by deduction) the special instance, or the concrete experience (by
induction) led to the recognition of the principle or law.
The analogy drawn, by bringing forward a fact supposedly
known, throws light on the phenomenon to be explained.
Analogy having this twofold reference and function, especially as it commonly takes the place of actual experimentation, illustrates in essence the inductive-deductive
procedure of science, which is not, as some would have us
believe, pecularily characteristic of modern times: it is the
natural behavior of the human mind, which needs only to
be duly exercised to result in a refined technique.   Moreover, in the passage quoted from Archytas, there is nothing
to show whether the parallels cited are to be regarded as
experiments or as observations casually made and recalled
as the law was being formulated; for it is in fact a law of
physics, although it is not stated with the mathematical
precision which we now require.

Illustrations of the procedure are to be found everywhere,
but certain medical treatises seem to be characterized by

a marked predilection for it. Thus the theory of the circulation of the fluids in the body,[10] including the respiratory and vascular systems, is explained by the reaction of heat and cold, respiration serving to cool the internal heat, whose seat is in the heart and veins. The heart especially attracts air (breath), as one may see by the action of a fire or burning torch in a closed chamber when there is no perceptible wind; for a draft is created which causes the flame to flicker. The Greeks explained a great number of phenomena by the attraction resulting from a (relative) vacuum caused by heat. This is particularly true of the theories of conception, as one may see by considering the various means proposed for promoting it in the Hippocratic writings. Aristotle herein follows the medical tradition when, speaking of the attraction exercised by the uterus because of its heat, he says:[11] "Hence it acts like cone-shaped[12] vessels which, when they have been washed with hot water, their mouths being turned downward, draw water into themselves." The cupping-glass is the standing illustration of attraction among the Hippocratics.[13] By this same process of attraction, or suction, is explained the nutrition of the body, in which like humor attracts its like from the stomach, as the sap of plants attracts its like from the soil, which contains all kinds of moisture.[14] Elsewhere we are assured that the human body is of a nature quite similar to plants.[15] Not only Theophrastus but the Hippocratics knew that plants

[10] Hippocrates, *De Natura Pueri*, 12 (VIII. 486 ff., L.); *De Carne*, 6 (VIII. 592 f., L.).

[11] *De Generatione Animal.*, II. 4, 739 b 11 ff., tr. Platt.

[12] A probable conjecture of Platt's.

[13] This is obvious, though it is mentioned only *De Prisca Medicina*, 22 (I. 626 f., L.).

[14] Hippocrates, *De Morbis*, IV. 34 (VII. 544 f., L.); cf. *De Humoribus*, 11 (V. 490 f., L.).

[15] Hippocrates, *De Natura Pueri*, 26 (VII. 528 L.).

differ considerably according to the character of the soil in which they grow, as our bodies are affected by different kinds of food.

Certain analogies of a particularly interesting sort are developed at length.   The supposed influence of the shape and size of the matrix upon the child is illustrated by what appears to be an experiment deliberately made.   We are told that if one takes a cucumber, after the blossom has passed, and places it, still young and attached to the vine, in a cup, it will shape itself to the container, if the latter be approximately of the right size.   "In general, one may say that all growing things are of such fashion as one forces upon them.   So also it is with the embryo: if it have sufficient room for growth, it becomes larger, if not, smaller."[16] Another analogy is found in the growth of trees whose roots are hampered by stones in the soil or by other restrictions of space.   The effect of heat and cold on the humors of the body is compared to the curdling of milk by fig-juice and the separation of the butter-fats from the whey in the Scythian process of churning.[17]   The formation of bladder-stones is explained as due to impurities in the milk imbibed by the child, hardening and granulating under the influence of phlegm in the urine.   At first a sediment of very small grains settles at the bottom of the bladder, as impurities in water settle after being stirred; then under the combined action of the raw phlegm it calcifies and grows, the process being compared to that of smelting impure iron-ore, in which after the fusing and removal of the slag the pure metal remains a solid mass in the furnace.[18]   A somewhat similar illustration serves to explain how the passage of air

[16] Hippocrates, *De Semine*, 9 (VII. 480 f., L.).
[17] Hippocrates, *De Morbis*, IV. 52 (VII. 590 L.); cf. *ibid.*, 51 (VII. 584 L.).
[18] *Ibid.*, IV. 55 (VII. 600 f., L.); cf. *De Aëre, Aquis, Locis*, 9 (II. 38 L.).

through the embryo brings about the accession of like to like and the consequent articulation of the body. Under the action of air (breath) all things, we are told,[19] separate kind by kind. If you attach a tube to a bladder, and after passing earth, sand, fine grains of lead, and water into it, blow through the tube, the contents will first be thoroughly mixed, but if you continue blowing, in time lead will go to lead, sand to sand, and earth to earth. Let it dry and tear the bladder, and you will find like gathered to like. So the embryo and the body are articulated and in them like joins like.

Again, certain phenomena connected with various kinds of congestion called for explanation, such as the accumulation of blood in diseased or injured tissues or vessels. The question naturally arose why the humor was not drained off, and the solution of the difficulty was found by comparison with a liquid contained in a narrow-necked vessel which has been inverted. The unquent in a lecythus will not flow out, if it be kept vertical, because it stops the aperture; but if the vessel be tilted to one side, the stoppage will cease and the unguent discharge itself. Once the flow begins it continues as in the case of "water on the table."[20] This latter phenomenon is again mentioned[21] and evidently attracted considerable attention. It is not fully described in the Hippocratic writings, and there seems to be some doubt as to what special effect is meant. The most probable conjecture is that reference is made to water standing in considerable quantity on a level table and not flowing off, despite the fact that its upper surface is appreciably higher than that of the table, until the liquid is in direct communication with the edge. Once a beginning is made, the whole

[19] Hippocrates, *De Natura Pueri*, 17 (VII. 498 L.).
[20] Hippocrates, *De Morbis*, IV. 51 (VII. 588 L.); cf. *ibid.*, 57 (VII. 612 L.)
[21] Hippocrates, *De Natura Pueri*, 18 (VII. 502 L.).

will follow.   The phenomenon is now explained as due to
surface-tension, which in certain liquids, like mercury, will
produce perfect spherical drops.   Occasionally one can not
restrain a smile at the analogies suggested, as when the
tendency to yawn, when a fever is coming on, is explained
by air collecting in quantity and forcing its way out at the
mouth which it pries open, as steam rising from a boiling
kettle will lift the lid.   In the same passage, however,
sweat is more reasonably compared to steam condensing on
the cover of a boiling kettle.[22]

Most of the examples just cited are derived from a group
of treatises that belong together and are probably parts
of a whole composed by a single writer who greatly affected
mechanical parallels to physiological processes.   Littré[23]
says of him, "He believes he understands the phenomena of
life, when he has found physical phenomena exactly corresponding to them."   One can not deny the scientific value
of some of these analogies, which at times have almost the
character of experiments.   At the close of the Fifth Century
B. C. there seems to have been a strong tendency[24] to make
intelligible obscure phenomena by bringing them into direct
connection with familiar things to be directly observed,
whether they occurred naturally or required to be artificially brought about.   In principle this procedure does not
differ from that which has been followed in recent times in
contriving mechanisms to illustrate and support a theory.

In the treatise *On the Old School of Medicine* occurs a
passage[25] which well illustrates analogical reasoning and is
worth quoting at length: "I hold that it is also necessary

[22] Hippocrates, *De Ventis*, 8 (VI. 102 f., L.)

[23] *Oeuvres Complètes d'Hippocrate*, VIII. 8.

[24] Other illustrations of this kind are to be found in Willy Theiler, *Zur
Geschichte der Teleologischen Naturbetrachtung bis auf Aristoteles*, 54–57.

[25] Hippocrates, *De Prisca Medicina*, 22–24 (I. 626 ff., L.), tr. Jones.

to know which diseased states arise from powers and which from structures. What I mean is roughly that a 'power' is an intensity and strength of the humors, while 'structures' are the conformations to be found in the human body, some of which are hollow, tapering from the wide to the narrow; some are extended, some hard and round, some broad and suspended, some stretched, some long, some close in texture, some spongy and porous. Now, which structure is best adapted to draw and attract to itself fluid from the rest of the body, the hollow and expanded, the hard and the round, or the hollow and tapering? I take it that the best adapted is the broad hollow that tapers. One should learn this thoroughly from unenclosed objects that can be seen. For example, if you open the mouth wide you will draw in no fluid; but if you protrude and contract it, compressing the lips, and then insert a tube, you can easily draw up any liquid you wish. Again, cupping instruments, which are broad and tapering, are so constructed on purpose to draw and attract blood from the flesh. There are many other instruments of a similar nature. Of the parts within the human frame, the bladder, the head, and the womb are of this structure. These obviously attract powerfully, and are always full of a fluid from without. Hollow and expanded parts are especially adapted for receiving fluid that has flowed into them, but are not so suited for attraction. Round solids will neither attract fluid nor receive it when it has flowed into them, for it would slip round and find no place on which to rest. Spongy, porous parts, like the spleen, lungs, and breasts, will drink up readily what is in contact with them, and these parts especially harden and enlarge on the addition of fluid. They will not be evacuated every day, as are bowels, where the fluid is inside, while the bowels themselves contain it externally; but when one of these parts drinks up the fluid and takes it to itself, the porous hollows, even the small ones, are everywhere filled, and the soft, porous part becomes hard and close, and neither digests nor discharges. This happens because of the nature of its structure. . . . . There are many other structural forms, both internal and external, which differ widely from one another with regard to the experiences of a patient and of a healthy subject, such as whether the head be large or small, the neck thin or thick, long or short, the bowels long or round, the chest and ribs broad or narrow, and there are very many other things, the differences between which must all be known, so that knowledge of the causes of each thing may ensure that the proper precautions are taken. . . . . If a man can in this way conduct with success inquiries outside the human body, he will always be able to select the very best treatment.''

It need hardly be said that not all analogies were drawn from inorganic nature. Aristotle compares the phenomena of earthquakes to the movements of the pulse, of trembling, or of spasms in the living organism,[26] and explains the flight of shooting-stars by adducing that of applepips propelled by being pressed between the fingers.[27] Though such examples of the processes in question may seem to us less illuminating, there is no real difference as regards evidential value; the significant point in regard to all analogies is that they illustrate the way in which the mind sets about its task of inferring the unknown from the known, of piercing the veil of mystery that hangs over all that lies beyond our ken, which constitutes the legitimate philosophical interest of the study of the history of science.

[26] *Meteorol.*, II. 8, 366 b 14 f.
[27] *Ibid.*, I. 4, 342 a 10.

V. EXPERIMENTATION

Modern science lays great stress on experimentation; and one often hears it said that herein it differs radically from ancient science. More discriminating judges find the essential difference in the fact that, while the Greeks did occasionally experiment, they did not develop and employ the experimental method, to which the triumphs of our times are attributed. If the former position is flagrantly at variance with the truth, it can not be said that the latter is either quite fair or intelligent. One may admit, and freely admit, that modern science is distinguished from much that anciently passed for such by the character and technique of its experimentation without allowing its claims to be essentially different in either character or technique. Those who have instituted a comparison have generally been scientists who were at home in the modern practice but had at best a superficial knowledge of ancient times. Few of their number seem to have given much thought to the logical and psychological aspects of the problem. There is no need of entering into an exhaustive consideration of the nature and function of experimentation, which may well be left to writers on the methodology of science; but the layman may properly require such a statement as will enable him to understand and appraise the procedure of the pioneers. The ideal structure, which we call science, is of course founded on experience, as is every ideal structure. What distinguishes it from others is the degree of intimacy it establishes with experience, which controls it and is in turn controlled by it. Man has always derived his notions from experience and has always appealed to experience for confirmation of his notions. In this "experience" both observation and experiment are from the first included.   Observation is the essential, but it may be immediate or remembered; in experimentation there is involved the intentional doing of something in order that one may observe what will occur, but the experiment itself may be past, present, or prospective.   A remembered experiment is merely a stored observation.   Experiment obviously has the advantage over mere observation that it may be repeated and varied, generally at will; but in the end it is the observation which gives value to the experiment.   This experience is so elementary that one can not think of human beings as without it; one sees the newborn child obtaining it as truly as the most ingenious scientist.   In evidential value, as in principle, there is no reason for giving experiment the preference over observation.   In some sciences, which are (or may be) quite as exact as the experimental, there is either no or little possibility of experimenting.

Experimentation thus being only a form of observation, its uses and functions are those of experience in general; and, as has already been said, experience serves us in forming and confirming (or correcting) our notions.   A scientist is apt to say that it supplies the facts and the means of testing the inferences from them.   Actually the "facts" are as truly constructs as the theories evolved from them; both are notions which require the test of experience.   The experimental method, therefore, is nothing more than an elaborate technique for proving the value of the facts, or data, and of the generalizations drawn from them.   Where the nature of the subject, or the degree of definition of the problem in hand, admits of the application of experimentation, it is clearly an advantage, because of the possibility of direct control, to resort to this method at every stage of induction and deduction, taking note of the phenomena in

the light of the implications involved in the process.   Ordi-
narily, the careful reasoner will hold himself absolved from
further tests if he finds that he has duly followed the rules
of logic; but the experimental checking of all conclusions
is not only preferable but also more primitive, being the
original procedure.

There is, therefore, nothing novel in experimentation, or
in the experimental method, as such.   What distinguishes
the best scientific procedure of modern times is chiefly the
refinement of technique, and, in a few outstanding scientists, the recognition of the methodological principles which
require an elaborate technique.   This refinement of technique is due principally to the progressive definition of
problems as science has pushed its inquiries farther and
farther.   The pioneers were for the most part occupied
with large and general questions, and their tools and methods were correspondingly crude and rough; but progress
consists generally in dividing and defining the question, and
each step forward calls for added precision.   One does not
require a microtome to cut wood for a camp-fire.   In methodology the chief improvement is doubtless due to the
failure of early attempts at generalization from "facts."
The object of generalization is to explain or control "facts;"
but if the attempt proves abortive, one is inevitably baffled
and scrutinizes the supposed facts.   This has no doubt always been done in practice, but theory was slow to recognize the problem.   Failure was laid to the charge of haste
or other heedlessness, but no technique of determining the
facts was developed.   Hence, while Greek science and logic
safeguarded the processes of induction and deduction, it
remained in general for modern science to recognize the need
of methods of ascertaining and defining the initial "facts"
and to apply experimentation to this end.   It is at this
point that one has to acknowledge a decided improvement

in modern times; it is not merely a question of technique, but, where its implications are comprehended, may be regarded as a matter of principle.

For reasons already stated the records of experimentation among the Greeks are clearly not representative of their practice. A modern scientific monograph describes in detail the apparatus and the method employed. This course is adopted in order to enable other workers to judge of their adequacy and, if they should so desire, to repeat the experiments. One has to go down almost to the close of ancient Greek history before one finds anything corresponding to this procedure. Even where an experiment is mentioned, it is usually done by the way, without detail, as one may refer to a well-known fact by way of illustration. The important point being that the phenomenon in question has been noticed, or may be noticed under certain circumstances, one is frequently left in doubt whether reference is made to a simple observation, to an observation repeatedly made, or to one obtained in the course of deliberate and intentional experiment. The main reason, however, for the infrequency of recorded experiments is the studied brevity of exposition affected by Greek scientists, who created the models still followed in most treatises. Conclusions are offered with only occasional citation of supporting evidence. One sees this practice exemplified to perfection in the published forensic orations. In the court the sworn testimony of the witnesses was presented in full at the appropriate moment; but it is not included in the speech, which gives at best a summary of it and generally merely indicates that there is substantiating evidence for the contention of the speaker. It is only a genius like Plato who can lead up to his conclusions, presenting the matter inductively, and yet satisfy the Greek's love of aesthetic form; but he was dealing for the most part with ethical notions instead of the "material

facts" that occupy the so-called exact sciences of our time.
It is interesting, nevertheless, to note that it was apparently
Plato who first clearly distinguished the exact sciences from
the empirical arts which use little or no mathematics.[1]

Herodotus, an author in many respects hardly representative of the Greek method of exposition, offers a number of examples of the way of presenting evidence in detail
for his statements. After recounting certain inventions and
discoveries of the Egyptians and reporting their claim to
have first recognized the Olympian gods, whom the Greeks
learned of from them, he says:[2] "They showed me most of
this by plain proof. The first human king of Egypt, they
said, was Min. In his time all Egypt save the Thebaic
province was a marsh; all the country that we now see was
then covered by water, north of the lake Mœris, which lake
is seven days' journey up the river from the sea. And I
think that their account of the country was true. For,
even if a man has not been told it, he can at once see, if he
have sense, that the Egypt to which the Greeks sail is land
acquired by the Egyptians, given them by the river—not
only the lower country but even all the land to three days'
voyage above the aforesaid lake, though the priests added
not this to what they said. For this is the nature of the
land of Egypt: firstly, when you approach to it from the
sea and are yet a day's run from land, if you then let down
a sounding-line you will bring up mud and find a depth of
eleven fathoms. This shows that the deposit from the land
reaches thus far . . . . . The greater portion, then, of this
country whereof I have spoken was (as the priests told me,
and I myself formed the same judgment) land acquired by
the Egyptians; all that lies between the ranges of moun-

[1] *Philebus*, 55 e ff.; cf. *Euthyphro*, 7 b ff.
[2] II. 4–5, 10–12, tr. Godley.

tains above Memphis seemed to me to have been once a
gulf of the sea, just as the country about Ilion and Teu-
thrania and Ephesus and the plain of the Maeander, to compare these small things with great. . . . . There are other
rivers, not so great as the Nile, that have wrought great
effects . . . . chief among them is Achelous, which, flowing
through Arcarnania and issuing into the sea, has already
made half of the Echinades islands to be mainland.   Now
in Arabia, not far from Egypt, there is a gulf of the sea entering in from the sea called Red. . . . . I hold that where
now is Egypt there was once another such gulf; one entered
from the northern sea toward Ethiopia, and the other, the
Arabian gulf of which I will speak, bore from the south toward Syria; the ends of these gulfs pierced into the country
near to each other, and but a little space of land divided
them.   Now if the Nile choose to turn his waters into this
Arabian gulf, what hinders that it be not silted up by his
stream in twenty thousand years?   Nay, I think ten thousand would suffice for it.   Is it then to be believed that in
the ages before my birth a gulf even much greater than this
could not be made into land by a river so great and so busy?
Therefore, as to Egypt, I believe those who so speak, and I
am myself fully so persuaded; for I have seen that Egypt
projects into the sea beyond the neighboring land, and
shells[3] are plain to view on the mountains, and the ground
is coated with salt (insomuch that the very pyramids are
wasted thereby), and the only sandy mountain in Egypt
is that which is above Memphis; moreover, Egypt is like
neither to the neighboring land of Arabia, nor to Libya, no,
nor to Syria . . . . it is a land of black and crumbling earth,

---

[3] Fossiliferous rocks abound in Greece; cf. Frazer, *Pausanias's Description of Greece*, II. 546. From the time of Xenophanes, in the Sixth Century
B.C., marine fossils found on lands attracted attention and were rightly
interpreted.

as if it were alluvial deposit carried down the river from
Ethiopia; but we know that the soil of Libya is redder and
somewhat sandy, and Arabia and Syria are lands rather of
clay and stones." Herodotus then goes[4] on to point out
that according to reports of Egyptian priests the level of
the Delta has risen within historical times, confirming the
conclusion that Egypt proper (the flood-plane of the Nile)
is the gift of the river.

Another example of the same sort is that relating to the
theories proposed by earlier Greeks to account for the
annual flood of the Nile.[5] One theory attributed the
inundation to the resistance of the etesian winds, which regularly blow from the northwest during the time of the overflow; the second derived the Nile from the river Oceanus,
which was supposed to flow round the inhabited earth; the
third explained the flood as due to the melting of snows
in equatorial Africa. Herodotus dismisses the two other
theories very summarily and addresses himself to the refu-
tation of the third:[6] "It has no more truth in it than the
others. According to this, the Nile flows from where snows
melt; but it flows from Libya through the midst of Ethiopia,
and issues out into Egypt; how then can it flow from snow,
seeing that it comes from the hottest places to lands that
are for the most part colder? nay, a man who can reason
about such matters will find his chief proof, that there is
no likelihood of the river's flowing from snow, in this, that
the winds blowing from Libya and Ethiopia are hot. And
the second proof is that the country is ever without rain
and frost; but after snow has fallen there must needs be
rain within five days;[7] so that, were there snow, there would

[4] II. 13.

[5] *Ibid.*, II. 20 ff.

[6] II. 22, tr. Godley.

[7] This looks like folk-lore. Aulus Gellius, VIII. 4, criticises the statement as an unwarranted generalization.

be rain in these lands.   And the third proof is, that the
men of the country are black by reason of the heat.   Moreover, kites and swallows live there all the year round, and
cranes, flying from the wintry weather of Scythia, come
every year to these places to winter there.   Now, were
there but the least fall of snow in this country through which
the Nile flows and whence it rises, none of these things
would happen, as necessity proves."   Having to his satisfaction disposed of other views he states his own, which
explains away the flood instead of explaining it; he holds
that the proximity of the sun, which is then far south, dries
up the sources of the Nile in winter, while in summer they
flow with their normal volume.

Of the two examples here cited Herodotus may properly
lay claim only to the second, which, though much admired
by some,[8] does him little credit; for he not only failed to appreciate the third theory at its true worth, but gave no indication that he comprehended the meaning of the other
two, which are founded on the old Ionian conception of the
dip of the earth-disk, causing a flow southward of water,
which was supposed to return from the southern ocean
in the flood of the Nile when the earth rights itself in summer.   The fluctuation of the ocean is followed by air-
currents[9] (trade-winds), of which the etesian winds are examples.   Herodotus, in saying that during the winter the
sun is driven southward by the storms,[10] showed himself

[8] For example, Whewell, *History of the Inductive Sciences*, I. 29 f., 33 f.

[9] Compare Hippocrates, *De Natura Pueri*, 25 (VII. 522 L.), and Herodotus, II. 19.

[10] II. 24; cf. Hippocrates, *De Ventis*, 2 (VI. 94 L.); Aëtius, II. 23 (Diels,
Doxographi Graeci, 532), for Anaximenes and Anaxagoras.   See also Otto
Gilbert, *Die Meteorologischen Theorien des Griechischen Altertums*, 490, nn.
1 and 2 (for Anaximenes and Anaxagoras) and 216, n. 2 (for Epicurus).
The theory has considerable ramifications, which can not be followed here.

indebted to the same view. The first example, which is one
of the best specimens of scientific reasoning in Greek literature, is derived from Hecataeus of Miletus. Both examples
are, however, exceptional in that they attempt to present
rather fully the evidence for the views adopted. There is
here no room for experimentation, but we have abundant
proof of close observation of conditions, especially in the
discussion of the geological history of Egypt.

Scientific method may be displayed not only in direct
induction and deduction, but quite as well, if not at times
more fruitfully, in criticism of false hypotheses. Aristotle
was, within limits, the ideal critic, because, as the first
codifier of logic, he was keenly observant of the almost infinite forms of false inference. In the method, which he
followed as a rule, of discussing the difficulties and problems
before he developed his own theory on a given subject, he
found a means of clearing the way for a fair survey. To
be sure, he often proceeded merely linguistically, pointing
out the various acceptations of terms; but this possessed
the virtue of indicating the necessary distinctions to avoid
ambiguity, even when it did not constitute a valid induction. His logical method was often extremely illuminating
in criticism, as for example in his discussion of the doctrine
of pangenesis,[11] of which Mr. Platt says,[12] "Aristotle's insight in all this passage is miraculous." Even here, however, Aristotle made mistakes; but one must say in all fair-
ness that in this case, as in many others, the fault lay, not
in his method, but in the means at the time available for
making observations. As a scientific achievement this es-
say in criticism ranks with the great discoveries.

But we are now to turn our attention to Greek examples
of experimentation. As one might expect from the pro-

[11] *De Generatione Animal.*, I. 17–18.
[12] In his note on 723 b 32.

cedure in practical life, there are abundant instances of trying this or that expedient in order to see what the result would be. It matters not at all whether the thing was first done accidentally or on purpose; if the consequence was desirable, one repeated the operation, which might or might not prove to yield the same satisfaction. The method of trial and error, with the gradual definition of the conditions necessary for success, is as nearly as possible fundamental, and at every step it is tentative or experimental. Herodotus[13] describes the way in which Cambyses "tried the spirit" of King Psammenitus by subjecting him to extreme mental torture—a course that one may observe followed by children.

Improvements in every practical pursuit rest on experiment; and the terms used to express the operation are naturally the same as those applied to scientific procedure. Plato[14] tells how the possessor of the ring of Gyges, sitting in the assembly of the shepherds, chanced to turn the collet of the ring inside his hand, when instantly he became invisible to the rest of the company and they began to speak of him as if he were no longer present. He was astonished at this, and again touching the ring he turned the collet outward and reappeared; he made several trials of the ring, and always with the same result—when he turned the collet inward he became invisible; when outward, he reappeared. Though fiction, the story presents the type, and shows how familiar it was to every one. Plato's supposed objection[15] to physical experiments may be regarded as resting, so far as it may be taken seriously, on the want of technique and a preference for the absolute demonstration of pure mathematics.

[13] III. 14.
[14] *Republic*, 359 e ff.
[15] *Republic*, 531 a; *Philebus*, 68 d.

The incident of the ring of Gyges does not differ in any essential way from confirmatory observation.   Thus Aulus Gellius[16] says that he wanted to determine whether, as was asserted, the leaves of the olive-tree turned their grey undersides upward on the day of the winter solstice, and that he found that it was generally true.   Of a piece with this are the soundings taken, according to Herodotus, in the Mediterranean as one approached the Delta of the Nile[17] and at the First Cataract, where between the hills Crophi and Mophi there were reported to be the bottomless springs from which issued the flood of the Nile.[18]   Herodotus scouts the story told by the recorder of the treasury of Athene at Sais about the test made by King Psammetichus, who was said to have let down a rope of many fathoms' length without finding bottom, and says, "If the recorder spoke truth, he showed, as I think, that here are strong eddies and an upward flow of water, and the rushing of the stream against the hills makes the sounding-line, when let down, unable to reach the bottom."   Here common-sense prevails over experiment.   An attempt to lay an experimental basis for history is delightfully told :[19] "Before Psammetichus became king of Egypt, the Egyptians deemed themselves to be the oldest nation on earth.   But ever since he desired to learn, on becoming king, what nation was oldest, they have considered that, though they came before all other nations, the Phrygians are older still.   Psammetichus, being nowise able to discover by enquiry what men had first come into being, devised a plan whereby he took two newborn children of common men and gave them to a shepherd to bring up with his flocks.   He gave charge that none should speak a word

[16] IX. 7, 1 f.
[17] II. 5.
[18] II. 28.
[19] Herodotus, II. 2, tr. Godley.

in their hearing; they were to lie by themselves in a lonely hut, and in due season the shepherd was to bring goats and give the children their milk and do all else needful.   Psammetichus did this, and gave this charge, because he desired to hear what speech would first break from the children, when they were past the age of indistinct babbling.   And he had his wish; for when the shepherd had done as he was bidden for two years, one day as he entered both the children ran to him stretching out their hands and calling 'Bekos.' When he first heard this he said nothing of it; but coming often and taking careful note, he was ever hearing the same word, till at last he told the matter to his master, and on command brought the children into the king's presence. Psammetichus heard them himself, and inquired to what language this word Bekos might belong; he found it to be a Phrygian word signifying bread.   Reasoning from this fact the Egyptians confessed that the Phrygians were older than they.   This is the story which I heard from the priests of Hephaestus's temple at Memphis; the Greeks relate (among many foolish tales) that Psyammetichus made the children to be reared by women whose tongues had been cut out."

The story is obviously an Ionian skit, most probably borrowed from the wily Hecataeus, who may be assumed to have had more knowledge of Phrygian than the priests of Memphis.   As a psychological experiment it is perfect; anyone who knew children and goats could predict the result, and a little knowledge of languages would supply the word and the interpretation of the infant bleatings.   The sly Greek could not refrain from a reference to the talkativeness of women, whose tongues would have to be cut out if the experiment was to succeed, while the shepherd might be trusted to hold his own.   One may safely conjecture that Herodotus has omitted one step of the reasoning; for the experiment would naturally be understood to show the

natural, as opposed to conventional, language; for the
Greeks were fond of looking to animals for evidence of
what was natural.[20]

Many illustrations used by writers imply or recommend
simple experiments. Thus the Hippocratic treatise *On Airs,
Waters, and Localities*[21] recommends boiling water on the
ground that it serves to purify it in the same way that the
sun, in "drawing" it by evaporation, purifies even the water
from stagnant pools when it descends in rain.   In the same
way boiling honey is recommended,[22] because it removes
the impurities.   One often can not distinguish between
experiment and observation in such cases, as when Aristotle
says,[23] "If you take out of the shells a number of yolks and
a number of whites and pour them into a saucepan and boil
them slowly over a low fire, the yolks will gather into the
center and the whites will set all round them," and again,[24]
"If we enclose many eggs together in a bladder or something of the kind and boil them over a fire so as not to make
the movement of the heat quicker than the separation of
the white and the yolk in the eggs, then the same process
takes place in the whole mass of eggs as in a single egg, all
the yellow part coming into the middle and the white surrounding it."   A Hippocratic treatise[25] says that if one puts
water and oil in a cauldron and builds a large fire under it,
a great deal of the water, but little of the oil, will evaporate.
Mention is made[26] of the fact that blood coagulates and sur-
rounds itself with a membrane when it cools undisturbed,

[20] Cf. Herodotus, II. 64.   Many other passages might be cited.
[21] *De Aëre, Aquis, Locis*, 8 (II. 36 L.).
[22] Hippocrates, *De Victu in Acutis*, 15 (II. 348 L.).
[23] *Historia Animal.*, VI. 2, 560 a 30 f, tr. Thompson.
[24] *De Generatione Animal.*, III. 1, 752 a 4 ff., tr. Platt.
[25] *De Morbis*, IV. 49 (VII. 580 L.).
[26] Hippocrates, *De Carne*, 8–9 (VIII. 594 f., L.).

but does not if stirred; similarly we are told that the liquid portion of the eye is viscous: "We have often observed the viscous liquid issuing when an eye is broken open; if it is still warm, it is liquid; but when it cools it becomes dry, like transparent incense—the same in man as in animals."[27] The transparent object here meant is presumably the crystalline lens which may have been observed surrounded by the vitreous humor and the hyaloid membrane. The remark that the same is true of the animal-eye suggests that the latter was experimented on for comparison and confirmation. We are informed that hairs will not grow if one burns the epidermis (destroying the dermic papillae) and produces a blister,[28] and that pressure produces heat and may cause combustion.[29] Lactation, or the secretion of milk by a woman with child, is explained by the pressure of the embryo on the surrounding parts, forcing the fatty ingredients of the food and drink into the omentum and the flesh, "just as if one were to oil leather and let the oil sink in, and when it was absorbed should squeeze out the leather: when it was pressed, the oil would ooze out."[30] The author evidently is quite familiar with transudation or osmosis. An observation, probably often repeated experimentally, is attributed to Anaximenes, the Ionian philosopher: if one breathes with mouth wide open, the breath is warm; if the mouth is nearly closed, it is cold.[31] That is true, judged by sensation alone; one must remember that

[27] *Ibid.*, 17 (VIII. 606 L.).

[28] Hippocrates, *De Natura Pueri*, 20 (VII. 508 L.).

[29] *Ibid.*, 24 (VII. 520 L.).

[30] *Ibid.*, 21 (VII. 512 L.). Cf. Herodotus, IV. 2 for the Scythian procedure in milking mares. The supposed action of pressure, in the latter case the pressure of air, is the reverse of that assumed in the Pseudo-Aristotelian *Problems*, II. 1, 866 b 9 ff., mentioned below, p. 176.

[31] Plutarch, *De Primo Frigido*, 7, p. 947 f.; cf. Pseudo-Aristotle, *Problem.* 34, 7, 964 a 10 ff.

the thermometer had not been invented.   The pardonable
mistake explains the oft-repeated assertion that springs are
warm in winter and cold in summer,[32] a phenomenon which
was variously explained.   Aristotle asserted, while others
denied, that an inflated bag weighs more than an empty
one.[33]   Whether any container he possessed would with-
stand sufficient pressure, or any balance available was deli-
cate enough to enable him to justify his statement, is more
than doubtful; but there can be no question that he tried to
prove the point by actual weighing.   So we are told that
various philosophers used inflated bags as well as the water-
clock as proof that air is a substance and not a void.[34]   "In
the case of plants," says Aristotle,[35] "there are some that
are observed to live when they are divided and the parts
are separated from each other, showing that there is in each
of these plants in actuality a single soul, but potentially
several souls.   The same thing we observe in different varieties of soul, as in the case of insects that have been dismem-
bered."[36]   In trying to explain the humming note of certain
insects[37] he expresses the notion that it is due to the air,
which exists internally, causing a rising and falling move-

[32] Oenopides, in Diodorus Siculus, I. 41, 1; Hippocrates, De Natura Pueri,
24–25 (VII. 518 ff., L.), De Aëre, Aquis, Locis, 3 (II. 16 L.), 7 (II. 30 L.);
Herodotus, IV. 181; Theophrastus, De Igne, 16; Apollonius Rhodius,
Argonautica, III. 225 f.; Pausanias, VIII. 28, 2–3; Seneca, Natur. Quaest.,
IV. 2; Athenaeus, Deipnos., II. 87; Galen, XI. 555, ed. Kühn.   The Jews
also shared this view: cf. Pesachim, IX. tr. Rodkinson; Jewish Encyclo-
paedia, VIII. 679 a; Aristotle, Meteorol., I. 12, 348 b 30 ff., expressed the
belief that water freezes more quickly if previously heated.   The change of
temperature would of course be more marked, and water does freeze more
readily if the contained air is driven off by heating.
[33] De Caelo, IV. 4, 311 b 8 ff.
[34] Aristotle, Physica, IV. 6, 213 a 22 ff.
[35] De Anima, II. 9, 413 b 16 ff.
[36] Cf. Aristotle, De Respiratione, 3, 471 b 19 ff.
[37] Ibid., 9, 475 a 9 ff.

ment like that produced by breathing in the thorax of animals that have lungs: "What occurs is comparable to the
suffocation of a respiring animal by holding its mouth, for
then the lung causes a heaving motion of this kind. . . . . It
is by friction against the membrane that they produce the
humming sound, as we said, in the way that children do
by blowing through the holes of a reed covered by a thin
membrane." The present-day equivalent of this favorite
instrument of children is a rubber band stretched across
one end of a spool. In all these cases the observations
were either the result of experimentation or prompted experiments to confirm them in the interest of theory. In
other instances the value of the observations is virtually
that of an experiment, as when Aristotle reports as follows
about the fertility and rapid maturity of mice:[38] "On one
occasion a she-mouse in a state of pregnancy was shut up
by accident in a jar containing millet-seed, and after a little
while the lid of the jar was removed and upward of one
hundred and twenty mice were found inside it." One won-
ders whether Mr. Ellis P. Butler got the idea of his extrav-
aganza *Pigs Is Pigs* by reading this passage.

Greek writers abound in similar observations and simple
experiments, showing how familiar the practice of experimentation was to them. Aristophanes[39] represents the
school of Socrates as experimenting with a flea to determine
how many of its feet it could jump. The jest proves that
experiments involving measurements were far from unknown in Fifth Century Athens. Pliny the Elder records
the marking of a dolphin's tail (as one bands birds today for
various purposes) in order to ascertain how long it lived,
if it should ever chance to be captured again, and says that
this chance befell after three hundred years;[40] and mentions

[38] *Historia Animal.*, VI. 37, 580 b 10 ff., tr. Thompson.
[39] *Nubes*, 144 ff.
[40] *Naturalis Historia*, IX. 7.

the case of a man at Rome being cast into a pit with ser-
pents to determine whether he was immune from their
stings.[41] One can not think that the interest in this case was
scientific, for less cruel means might easily have been found
to decide the question. When he says[42] that the well at
Syene, where the sun was proved to shine vertically into
its depths at the summer solstice (giving Eratosthenes the
necessary datum for the calculation of the earth's circumference), was sunk for the purpose of the experiment, he
reveals how common even elaborate experiments were in
his time; but there is every reason to think that he was
mistaken, for the vertical gnomon of a sun-dial, which cast
no shadow, would serve equally well and make the laborious
undertaking ridiculous. In discussing the question whether
fishes breathe, Aristotle[43] reveals that he had experimented
with them, as well as with tortoises and frogs, which were
forcibly kept under water until they drowned. The Hippocratics were aware of the sublimation, or the immediate
vaporization of ice without passing through the liquid
phase. "All waters derived from snow and ice are bad; for
once they freeze, they do not return to their former character, but the bright, soft, and sweet part is separated off and
disappears, leaving the more turbid and heavy. You may
observe this in the following way. In winter, if you please,
pour water by measure into a vessel and place it under the
sky where it will best freeze; on the following day bring it
into a warm room, where it will most quickly melt. When
it melts, measure the water again, and you will find its
quantity much less."[44] With this one may compare a
curious experiment suggested by Aristotle: "Of testaceans,"

[41] *Ibid.*, XXXVIII. 6.
[42] *Ibid.*, II. 75.
[43] *De Respiratione*, 3, 471 a 31 ff.; 9, 475 a 9 ff.
[44] *De Aëre, Aquis, Locis*, 8 (II. 36 L.).

he says,[45] "some that are incapable of motion subsist on fresh water, for, as the sea-water dissolves into its constituents, the fresh water from its greater thinness percolates through the grosser parts; in fact, they live on fresh water just as they were originally engendered from the same. Now, that fresh water is contained in the sea and can be strained off from it can be proved in a thoroughly practical way. Take a thin vessel of moulded wax, attach a cord to it, and let it down quite empty into the sea; in twenty-four hours it will be found to contain a quantity of water, and the water will be fresh and drinkable."[46] He mentions the matter elsewhere also,[47] and other authors repeat his statement without referring to him.[48] If the experiment was carried out with the care he recommends, Aristotle was deceived; possibly the drops of fresh water found in the wax vessel were due to the condensation of the moisture in the inclosed air.

In the latter passage Aristotle states that salt water is both thicker and heavier than fresh water, as is proved by its weight. From an early time the question was raised why the sea is salt, and the general answer was that the salt was obtained by water filtering through the earth. If that were true, it was natural to think that it could be removed by the process of filtering, as Aristotle suggests. The Greeks knew that a ship loaded to the gunwales might float in salt but sink in fresh water, as they were aware that

---

[45] *Historia Animal.*, VIII. 2, 590 a 18 ff., tr. Thompson. On this experiment, see Diels, *Hermes*, XL. 310 ff., and Otto Gilbert, *Die Meteorologischen Theorien des Griechischen Altertums*, 424, n. 2.

[46] Maine lobstermen assured me that the glass globes, which they use to buoy the ropes attached to their pots, when opened by breakage, are found to contain fresh water. I have been repeatedly asked to account for the fact.

[47] *Meteorol.*, II. 3, 358 b 35 ff.

[48] Aelian, *Historia Animal.*, IX. 64; Pliny, *Naturalis Historia*, XXXI. 37.

eggs will float in heavy brine.[49]   Aristotle repeatedly refers[50] to the fact that salt water is heavier than fresh, and he was not the first to test it by the use of scales.   In the Hippocratic treatise *On Airs, Waters, and Localities*[51] we are told that one must bear in mind the properties of various waters, since they differ in taste and weight; and later in the same work[52] the sun is said to draw up the lighter and sweeter parts of water, as one may see by the example of salt.   Theophrastus[53] told of the water near the Pangaean mines, which in winter weighed ninety-six drams per cup, in summer but forty-six.   Wonderful springs were thought to exist with water so light that nothing would float on them.[54]   In a most interesting passage Aristotle[55] explains that, owing to the nature of the elements, the same bodies do not seem to have the same weight, as, for example, a block of wood weighing a talent will be heavier in the air than one of lead which weighs a mina (= 1/60 talent), but will appear lighter in water.   In these and many other similar observations which might be mentioned, we have at least suggestions of experiments that paved the way for Archimedes' law of floating bodies and for the definition of specific gravity. Equally competent scientists differ on the question whether Archimedes may be said to have discovered specific gravity. It is apparently true that the technical term did not come into use until much later times; but the question really

[49] Aristotle, *Meteorol.*, II. 3, 359 a 12.   Cf. Ideler's note *ad loc.* for other references and for experiments with brine.   Galen, XI. 691, ed. Kühn.

[50] *Meteorol.*, II. 2, 355 b 32 ff.; Pseudo-Aristotle, *Problem.* XXIII. 8–9, 932 b 22, XXIII. 22, 934 a 9.

[51] *De Aëre, Aquis, Locis*, 1 (II. 12 L.); cf. Pliny, *Naturalis Historia*, XXXI. 38.

[52] 8 (II. 32 L.).

[53] Fr. clix, Wimmer.

[54] Strabo, XV. 703.

[55] *De Caelo*, IV. 4, 311 b 1 ff.

resolves itself into one of definition.   That the essential facts
were known to Archimedes is beyond dispute, though he
did not adopt a single definite standard to which all weights
were to be referred, as specific gravity is now guaged by
pure water at 4° centigrade.   Wiedemann[56] has called attention to a theorem (contained only in the Arabic text) pre-
fixed to the treatise of Archimedes which runs as follows:
"There are solid and liquid bodies, some of which are heavier
than others.   One body or one liquid is said to be heavier
than another, if, when one takes and weighs an equal
volume of each, one finds that one is heavier than the other.
If their weights are equal one does not say that one is
heavier than the other."   That we have here a statement, based on experiment, of the essential notion of specific gravity admits of no question.   The entire body of
doctrine concerning floating bodies, indeed, is obviously
based upon experimentation.   Moreover, the baryllium
which Bishop Synesius of Ptolemais, in the Fifth Century
of our era, by letter[57] requested Hypatia to procure for him
is conclusive evidence that the principle of specific gravity
was well understood.   Synesius calls it a *hydroscopium*; we
should call it a hydrometer.   It consisted of a tube with a
graduated scale, on which one could read the specific gravity of the samples of water offered.   The bishop was ill and
needed to exercise care in the water he drank.   His specifications for the hydrometer, however, are so indefinite that
one must conclude that he expected Hypatia, and perhaps
the instrument-maker at Alexandria, to be familiar with the
device.   In view of the familiarity of Greek physicians for
a millenium past with the differences in weight between different waters it would seem strange if the hydrometer was

[56] E. Wiedemann, "Über das Experiment im Altertum und Mittelalter,"
Unterrichtsblätter für Mathematik und Naturwissenschaften, XII (1906), 98 f.
   [57] *Epist.*, XV.

not long in use in ordinary practice. With these observations we may connect those which have to do with the circulation of water in communicating vessels. Empedocles is said[58] to have accounted in this way for the hot springs of Sicily, illustrating the process by the arrangements used in heating water for baths. One might incline to the assumption that the illustration was suggested by Seneca's knowledge of Roman baths;[59] but Plato in the *Phaedo*[60] clearly had in mind the same idea of the circulation of the ground-water between various reservoirs, although he apparently conceived its fluctuations as caused by the seasonal dip and partial righting of the earth itself. In one of the Hippocratic treatises[61] we find a specially clear statement of the facts of communicating tubes: "The reservoirs which I mentioned always irrigate the body, when they are full; when they become empty, they draw from it. The abdomen acts in the same way. It is as if one were to pour water into three or more cauldrons, set on a level and conveniently arranged, after fitting tubes into the holes; if one slowly pours water into one of the cauldrons, all the cauldrons will be filled from the one. If, when they are full, one draws from one of them, they will all give up the water again and empty themselves just as they were filled. It is just so in the body." Whether the apparatus was constructed for experimental or for practical purposes makes no difference.

With this contrivance and the scientific use of it one may compare that of the clepsydra, or water-clock, which is said to have been known to the Babylonians.[62] In principle it

---

[58] Seneca, *Quaest. Natural.*, III. 24.
[59] *Ibid.*, IV. 2, 28.
[60] 111 d.
[61] *De Morbis*, IV. 39 (VII. 556 f., L.).
[62] Gerland, *Geschichte der Physik*, 11.

was similar to the hour-glass, in which the flow of sand measured the time elapsed.   Of its practical uses we need not speak, except perhaps to mention the fact that the famous experimental physician Herophilus had a small one constructed with which he timed the pulse of his patients.[63]   In scientific theory the instrument played a considerable rôle by suggesting explanations of various phenomena.   One can imagine a Greek contemplating the action of this familiar device and drawing far-reaching conclusions from it as Newton did from the falling apple and Archimedes from the overflow of his bath.   Aristotle[64] adduces the support of a volume of water by the air in a clepsydra as illustrating the supposed action of the air in sustaining the weight of the earth, in reporting the theories of Anaximenes, Anaxagoras, and Democritus; but this may be his own suggestion, so far as concerns Anaximenes.   From the Pseudo-Aristotelian *Problems*[65] we learn, however, that Anaxagoras gave a detailed explanation of the water-clock, with which the later scientist was in general accord.   The theory is obviously based on extensive experimentation.   The reported title of a treatise by Democritus suggests that he may have proposed timing races with the clepsydra, or, according to a conjecture of Diels, discussed the correction of the practical time-piece by astronomical observations.[66]   If the latter be true, we might regard Democritus as a precursor of Hipparchus, who, as has already been suggested, must have used the clepsydra in determining latitude.   For such purposes the time-pieces then in use were not in themselves adequate.   In order to be serviceable there must be at

---

[63] Diels, *Antike Technik*[2], 27.

[64] *De Caelo*, II. 13, 294 b 13 ff.

[65] XVI. 8, 914 b 9 ff.

[66] Democritus, fr. 14[a], Diels, and Diels's note.   But the authenticity of the treatise is open to question.

least two having approximately the same rate and placed
at definitely measured distances apart.   Comparison of the
local time of lunar eclipses (which Hipparchus was the first
to utilize for this purpose), observed and recorded at different places, would at once give the difference in longitude.
Perhaps the most interesting use of the water-clock, however, is that made by Empedocles in illustrating his theory
of the circulatory system, which in his view included the
respiratory as well as the vascular.   Aristotle, who reports
his statement, says:[67] "He declares that inspiration and
expiration are due to the existence of certain vessels in
which there is blood, though they are not filled by it, and
they have passages into the outer air, smaller than the parts
of the body but larger than those of the air.   Now, since
the blood naturally moves up and down (*i.e.* outward from
the heart and back again), when it moves downward the
air enters and brings about inspiration, when it moves upward the breath rushes out and causes expiration.   He compares the process to that which one observes in the clepsydra: 'Thus do all things draw breath and breathe it out
again.   All have bloodless tubes of flesh extended over the
surface of their bodies; and at the mouths of these the outermost surface is perforated all over with pores closely packed
together, so as to keep in the blood while a free passage is
cut for the air to pass through.   Then when the thin blood
recedes from these, the bubbling air rushes in with an im-
petuous surge; and when the blood runs back, it is breathed
out again.   Just as when a girl, playing with a water-clock
of shining brass, puts the orifice of the pipe upon her comely
hand, and dips the water-clock into the yielding mass of
silvery water—the stream does not then flow into the vessel,
but the bulk of the air inside, pressing upon the close-

[67] *De Respiratione*, 7, 473 b 1 ff.

packed perforations, keeps it out until she uncovers the compressed stream; but then air escapes and an equal volume of water flows in—just in the same way, when water occupies the depths of the brazen vessel and the opening and passage is stopped by the human hand, the air outside, striving to get in, holds the water back at the gates of the ill-sounding neck, pressing upon its surface, till she lets go with her hand.   Then, on the contrary, just in the opposite way to what happened before, the wind rushes in and an equal volume of water runs out to make room.   Even so, when the thin blood that surges through the limbs rushes backward to the interior, straightway the stream of air comes in with a rushing swell; but when the blood runs back the air breathes out again in equal quantity.'"   Empedocles here propounds a theory of respiration which he shared with other thinkers, who thought of it as a species of nutrition, as it is in some low forms of animal life; for it includes perspiration through the pores of the skin.   The Pseudo-Aristotelian *Problems*[68] definitely connect sweat with the circulatory process which has just been described. The question is raised why one doesn't sweat when one forcibly holds one's breath, but rather when one expires freely; and the suggestion is offered that the breath, when held, may fill the vessels and so present the passage of sweat, as is the case with the water in a full water-clock, if one closes the openings.   In the Third Century B.C. Erasistratus sought by experiment to measure the amount of the insensible perspiration, by placing fowls in vessels and keeping them there without food for some time and weighing them at the beginning and end of the period.   Including the excreta at the end, he found that the loss of weight was considerable.[69]

[68] II. 1, 866 b 9 ff.   See above, p. 166, n. 4.
[69] Anonymus Londinensis, 33, 43 ff., p. 62 f., Diels.

We may next speak briefly of nutrition, which depends altogether upon experience. How much experimenting was done by primitive man in this matter and with how fatal results one can only guess, though it is reasonable to suppose that mere observation of other men and of animals was resorted to in general. Even now, it is said, swine are used to test the edibility of mushrooms and truffles. We have already quoted the remarks of a Hippocratic[70] on the experiences which made necessary the development of medicine as a science of dietetics. In the Fifth Century, Herodicus of Selymbria founded a school which relied upon dietetics and gymnastics for the cure of physical ailments. Iamblichus, in his *Life of Pythagoras*,[71] asserts that Pythagoreans first tried to determine the proper correlation between physical exertion, food, and rest; but since later writers were disposed to attribute almost every improvement to Pythagoras or his followers, one need pay little heed to their claims. Nevertheless, efforts were certainly made to establish such a correlation, and they were necessarily experimental. The Hippocratics,[72] however, were not content with generalizations, and endeavored to discover what foods were most readily digested by the individual, and to that end resorted to experiment, causing the patient to disgorge the contents of the stomach after a certain interval of time and examining the state of the several foods he had taken.[73] Galen and his father Nico instituted elaborate investigations and experiments with different foods, and foods differently prepared, to discover their nutritive value.[74]

[70] *De Victu*, I. 2 (VI. 470 L.), III. 67 (VI. 592 L.).

[71] 163.

[72] Suidas says that Hippocrates was a pupil of Herodicus; there can be no doubt that the Hippocratic treatise *De Victu* is indebted to him.

[73] Littré, *Oeuvres Complètes d'Hippocrate*, VI. 527.

[74] Galen, VI. 783 ff., ed. Kühn.

Nico, we are informed,[75] likewise experimented with wheat, barley, and other grains to ascertain whether it was true, as was commonly believed, that the seeds, even when carefully cleaned and selected, would on being planted change into darnel. He became convinced that the report was true, presumably because he had not sterilized the soil by heating it. One might in this connection mention other experiments made by physicians in the interest of diagnosis and prognosis.[76]

Dissection of course when practiced by scientists either is purely experimental or leads directly to experimentation. The Hippocratic treatise *On the Heart* reveals the contraction of that organ[77] and reports an experiment with the valves of the arteries.[78] The same work describes an experiment which was supposed to prove that, when one drinks, a certain quantity of water finds its way into the lungs: "The greater part of what a man drinks passes into the stomach; for the gullet is like a funnel and receives whatever we swallow; but a very small part, so small as not to be noticed, passes through the opening into the windpipe; for the epiglottis is an accurate cover and will not let anything more than a liquid pass through. Here you have the proof. If you mix water with a blue or red pigment and give it to a very thirsty animal—preferably to a pig, for the beast is neither nice nor squeamish—and if you then, while it is still drinking, cut its throat, you will find the windpipe discolored. But the experiment calls for an expert."[79]   One thinks of the means now adopted to test

[75] Galen, VI. 552, ed. Kühn.

[76] Hippocrates, *De Morbis*, II. 48 (VII. 72 L.), *Aphorism.*, V. 11 (IV. 536 L.); Aristotle, *De Generatione Animal.*, II. 7, 747 a 1 ff.

[77] *De Corde*, 8 (IX. 84 f., L.).

[78] *Ibid.*, 10 (IX. 86 L.).

[79] *Ibid.*, 2 (IX. 80 L.).

the adjustment of pistons of internal combustion engines. The experiment, though intended to be carefully controlled, led to a false inference, which was, however, challenged by other Hippocratics and by later writers.[80] The experimental attitude of the early physicians may be illustrated by several examples. One suggests that when one is dealing with an unknown ailment, it is well to administer a physic, not very strong; if the patient's condition improves, the proper course of treatment is indicated—one must continue to reduce and attenuate the system; if not, one must follow the opposite course.[81] Exploratory methods in acute pulmonary dropsy are elsewhere recommended, including, besides auscultation, the resort to incisions and the noting of the nature of the fluid drained off while carefully staunching the wound; in this way the character of the malady may be determined.[82] In connection with auscultation one should mention also the practice of resorting to succussion, generally to aid in diagnosis or, more specifically, to locate the collection of pus in a pleural sac,[83] less commonly in dealing with an effusion in the abdominal cavity,[84] where its effectiveness was questioned.[85] The general principle of experimentation is well expressed in the Hippocratic treatise *On the Art of Medicine*,[86] to which reference has already

[80] Hippocrates, *De Morbis*, IV. 56 (VII. 604 L.); Aulus Gellius, XVII. 11; Plutarch, *Quaest. Conviv.*, VII. 1, 1, 3.

[81] Hippocrates, *De Locis in Homine*, 34 (VI. 326 f., L.).

[82] Hippocrates, *De Morbis*, II 61 (VII. 94 f., L.).

[83] Hippocrates, *Praenotiones Coacae*, 424 (V. 680 L.); *De Morbis*, I. 6, 15 (VI. 150, 164 L.); *De Morbis*, II. 47 (VII. 70 L.); *De Morbis*, III. 16 (VII. 152 L.); *De Locis in Homine*, 14 (VI. 608 L.); *De Affectionibus Internis*, 23 (VII. 226 L.).

[84] Hippocrates, *De Locis in Homine*, 14 (VI. 306 L.).

[85] Hippocrates *De Morbis*, I. 17 (VI. 170 L.).

[86] *De Arte*, 12 (VI. 22 f., L.); cf. Daremberg, *Histoire des Sciences Médicales*, I. 131.

been made.   Of the eminent anatomists Erasistratus and Herophilus, who made Alexandria illustrious in the annals of medicine, it is not necessary to speak in detail, since their entire procedure was experimental; but it is well to remember that their example was followed by others.   In particular, Galen in certain fields, such as the physiology and anatomy of the sensory and motor nerves, adopted a distinctly experimental method, which was neither resumed nor improved until the time of Harvey.   Such is the judgment of the most competent historian of medicine;[87] and another[88] says that his investigation, by vivisection of animals, of the functions of the brain, the spinal cord, and the nerves entitles him to be called the founder of experimental physiology.

In embryology also the Greeks from early times made experiments.   Of the experiments and observations on the human embryo it is not necessary to speak, although they manifest a keen interest and a truly scientific spirit.[89]   One case observed in consequence of an abortion has greatly interested modern gynecologists, though it was wrongly interpreted.   There is, however, clear evidence of the scientific procedure in one of the Hippocratic treatises:[90]   "Take twenty or more eggs and set them under two or more hens to hatch; then, beginning with the second day, until the last, on which it is hatched, taking one each day, open and examine it, and you will find that all agrees with my account, as far as one may compare a chick with a man."   Not to speak of later writers, a careful study of the embryology of the chick is to be found also in Aristotle.[91]

[87] Daremberg, *ibid.*, I. 224–228.

[88] Haeser, *Lehrbuch der Geschichte der Medicin*, I. 364.

[89] Hippocrates, *De Carne*, 19 (VIII. 608 f., L.); *De Natura Pueri*, 13 (VII. 488 ff., L.).   Cf. Littré, *Oeuvres Complètes d'Hippocrate*, VII. 463 f., VIII. 577 ff.

[90] *De Natura Pueri*, 29 (VII. 530 L.).

[91] *Historia Animal.*, VI. 3.

The phenomena of capillarity were early known to the
Greeks from the use of lamp-wicks, which consisted of
various fibers, as of fibrous stalks.   This was not, however,
the extent of their knowledge.   There is a passage in Plato's
Banquet[92] which shows another application.   Socrates,
though expected, appeared late at the feast of Agathon,
having halted by the way to pursue the thread of some
thought that had occurred to him.   Upon his entrance he
was hailed by his host: "Come here, Socrates; recline by my
side, that I may touch and partake of the wise insight that
came to you in the porch; for of course you found and have
it—else you would not have desisted.   Socrates sat down
and said. 'It were well, Agathon, if wisdom were of such
sort as to flow from the fuller of us into the emptier, if we
were brought into contact, as water in cups flows through
the wool from the fuller into the emptier.   If that held true
of wisdom, I should greatly value reclining by your side;
for I think I should be filled with much fine wisdom.'"   If
the remark of Agathon suggests to our minds the virtue
that Jesus felt going forth from him, like a galvanic discharge, when he was touched by the unfortunate woman,[93]
Socrates clearly refers to siphoning off a liquid by capillary
action from one vessel into another.   The Greeks were well
aware also of the fact that wool, suspended in humid atmosphere, collects moisture[94] in quantities proportional to
the looseness of its texture.   One is not surprised, therefore, to find that the Hippocratics availed themselves of
it to explain certain physiological processes.   We are frequently reminded by them that the flesh of men is more
dense than women's; and in a notable passage this is cited
in explanation of the freer passage of the humors (and milk)

[92] *Symposium*, 175 c-e.
[93] *Mark*, V. 27–30.
[94] Lucretius, VI. 504.

through the flesh of females.[95]   "It is, as I have said before. I assert that woman's flesh is less dense and softer than man's; that being so, a women's body draws moisture from the abdomen more quickly and in larger quantity than a man's.   I say so, because if you place over water or a damp place for forty-eight hours scoured wool and a woolen garment scoured and closely woven, taking care that they be of the same weight, when you take them up and weigh them you will find the wool far heavier than the garment."   Many diagnostic and prognostic procedures, as for example those intended to discover whether a woman might be able to conceive, were based upon this assumption of the porosity of the female body; and the experiment with wool above suggested was turned to practical account in testing places in order to find the one best suited for digging a well.[96]

The discovery of the concordant intervals of the musical scale was early made by the Greeks, and is perhaps rightly attributed to Pythagoras.   Their existence is indeed readily detected by the ear alone, but their mathematical relations could obviously not be so discovered, as we know they were.   This achievement was possible only by experiment; but we have no certain record regarding the means and methods employed.   Everything relating to Pythagoras and the early Pythagoreans is involved in a tangled web of myth and fable, which doubtless contains much truth but does not admit of being reduced to an ordered story deserving of implicit confidence.   It would serve no useful purpose to recount the various tales of later date which purport to tell how Pythagoras made this momentous discovery; for most of them are inventions betraying by their inherent impossibilities an ignorance of acoustics on the part

[95] Hippocrates, *De Prisca Medicina*, 22 (I. 626 f., L.).

[96] Cf. Oder, "Quellensucher im Altertum," *Philologus*, Supplementband, VII. (1899).

of those who told them.   One version alone possesses a reasonable probability, because we know that the instrument
in question existed and that by the close observation of its
operation the numerical relations of the octave, the fifth,
and the fourth, attributed to Pythagoras, could be determined.   This version states that it was by the use of the
monochord that the truth was discovered.   Pythagoras
himself was credited with the invention of the instrument,
which throughout the Middle Ages served for the fundamental instruction in music.   With this simple device it
was readily possible to show the length of string required
to produce the concordant notes, the thickness and tension
of the string remaining the same.   The proportions $12:8:6$
could thus be established.   The discovery was momentous:
not only did it give a great impetus to the study of pure
mathematics, but it was an earnest of many applications of
mathematics to phenomena which at first sight appear to
defy analysis.   Of later developments of the theory of
sound as propagated by undulations of the air we need not
speak, except to remark that the analogy of waves in water
was recognized and the essential difference was noted that
sound travels in all directions.

The scientific study of optics also began early.   The
rudiments, at least, of perspective appear to have been
learned in the course of the Fifth Century B.C.   Vitruvius says[97] that "Agatharchus, in Athens, when Aeschylus
was bringing out a play, painted a scene, and left a commentary about it.   This led Democritus and Anaxagoras to
write on the same subject, showing how, given a center in a
definite place, the lines should naturally correspond with
due regard to the point of sight and the divergence of the
visual rays, so that by this deception a faithful representa-

[97] VII, *Praefatio*, 11, tr. Morgan.

tion of the appearance of buildings might be given in painted scenery, and so that, though all is drawn on a vertical flat façade, some parts may seem to be withdrawing into the background and others may be standing out in front." This statement, like the other of Vitruvius[98] relating to the experiments in acoustics tried in Greek theaters, has been drawn in question, because the evidence of contemporary vase-painting gives it no support; but this is by no means conclusive. When one considers the nice adjustments made in Greek temples, by substituting curved for straight lines,[99] in order to satisfy the eye, one must recognize the remarkable keenness of observation and patience in experiment necessary to attain the desired effect; and it is hardly conceivable that the architects of the Fifth Century depended solely on empirical tests without theoretical analysis. Among the works attributed to Euclid are treatises on *Optics* and *Catoptrics*, which of course rest on a foundation of observation and experiment. They establish the principles of optics, assuming that a ray of light follows a straight line, and treat of the images reflected in plane and spherical surfaces. The *Catoptrics* is now regarded as the work of a later author. It mentions a number of experiments, notably one which has with probability been ascribed to Archimedes: if an object is placed in a vessel so that it does not lie in the line of vision, it may become visible if water is poured into the vessel. This phenomenon is not explained in Euclid; but in Ptolemy's *Optics* the refraction of light is duly explained and the effects are carefully measured for different angles with an instrument devised for the purpose. The equality of the angles of incidence and reflection is shown; and the true values were ascertained for the angle of refraction both

[98] V. 3, 8.
[99] Cf. Marquand, *Greek Architecture*, 58.

when the ray passes from a rarer to a denser and from a
denser to a rarer medium.[100]   In this case we have a definite
statement of the results, and consequently modern scientists
have been pleased to recognize it as a valid experiment.

Aristotle has been severely criticized for his dictum regarding falling bodies.   He was certainly in error; but the
context is too often ignored.   He was discussing the doctrine which assumes an absolute void and was pointing out
difficulties which appeared to him to be insurmountable
on that view.   With his arguments we are not now concerned, but it should be borne in mind that he was occupied
with those problems and rejected the theory of an absolute
void.   We must remember also that he held the view that
certain things, e. g. the elements fire and air, had absolute
levity, naturally tending upward or away from the center,
while water and earth possessed absolute gravity, naturally
tending downward toward the center.   The Greeks had
considerable difficulty with a group of notions which are
correlates, such as warmer and colder, heavier and lighter,
larger and smaller, and which in their language were named
accordingly by compound, polar expressions, while we have
learned to use for each pair a single term—temperature,
weight, and size.   The acceptance of such terms, however,
depends on the acceptance of an arbitrary standard.   For
temperature, such a standard came with the invention of
the thermometer; for weight and measure, there were in
use standards, but only for practical purposes, and they
varied so greatly that an absolute standard for scientific
purposes was not as easily conceived or determined as one
may be tempted in our time to suppose.   A great deal
of the criticism of ancient science rests ultimately on the
tacit assumption that such standards ought to have been

[100] Cf. Gerland und Traumüller, *Geschichte der Physikalischen Experi-
mentierkunst*, 57.

used, when in fact they did not exist.   Thus, the question
whether the Greeks discovered the principle of specific
gravity is made to turn altogether on the point whether
they had, or had not, a standard, like pure water at 4°
centigrade, to which the weight of equal volumes could be
referred.   In this sense, but only in this sense, it is right to
say that the Greeks had not discovered it.   Perhaps the
most important achievement of modern science is the es-
tablishment of such arbitrary standards for all sorts of
phenomena; for it is by these means that the mathematical
formulation of natural laws becomes possible.

What Aristotle says, is this:[101] "We observe that things
having the greater momentum of gravity or levity, if they
be in other respects similar in form, move more quickly
through an equal space, in proportion to their magnitudes.
Consequently they must do so also through the void; but
that is impossible."   One suspects that the philosopher had
observed the fall of bodies of very different weight, the
lighter of which was noticeably retarded by the air; but in
any case his generalization was hasty.   Probably his notion
of absolute levity had something to do with his conclusion; for in speaking of the "momentum (rhopé) of gravity
or levity" he used a term which had intimate associations with "weighing."   Be that as it may, the dictum of
Aristotle regarding falling bodies has achieved an unenvi-
able notoriety because of the (apparently unfounded) tales
of Viviani regarding the hostility of the Aristotelians which
Galileo incurred by his experiments made from the leaning
tower at Pisa.   Whether Galileo actually made those
experiments is a question; but Simon Stevinus[102] in Holland
undoubtedly did formulate the laws after careful experimentation.   Not until recent years has it become known

[101] *Physica*, IV. 8, 216 a 13 ff.
[102] *Oeuvres*, II. 501.

that John Philoponus, a Platonist of the Sixth Century, discovered Aristotle's error by actual experiment.   I noted this fact nearly thirty years ago, supposing then that the observation of the studious Platonist was generally known : only when I found that Whewell and other historians ignored the point did I realize its historical significance. Meanwhile Wohlwill[103] had called attention to the passage in the commentary on Aristotle's *Physics*, in which Philoponus controverts the position of Aristotle :[104] "According to Aristotle," he says, "if the medium, through which the motion takes place, be the same, but the moving bodies differ in weight, their times must be proportional to their respective weights. . . . . but that is wholly false, as can be shown by experience more clearly than by logical demonstration.   For if you let two bodies of very different weight fall simultaneously from the same height, you will observe that the rate of motion does not follow their proportional weights, but there will be only a very slight difference in time, so that if their difference in weight be not very great, but one body were, say, twice as heavy as the other, the times will not perceptibly differ."

One of the primitive weapons, the sling, much employed in Greece, must have drawn attention at an early date to centrifugal force.   The sling consisted essentially of a strip of leather broad in the middle, where the missile was placed, and narrowing toward the ends, which were held in the hand.   By whirling it round and round, until it attained a maximum velocity, and then letting go one end of the thong, a shot could be hurled a long distance.   In the

[103] *Mitteilungen zur Geschichte der Medizin und der Naturwissenschaften,* (Leipzig, 1905), IV. 241.
[104] *Philoponi in Physicorum Libros Quinque Priores,* ed. Vitelli, 683, l. 7 ff.

Pseudo-Aristotelian *Mechanics*[105] an explanation of the effect is sought and found. Other phenomena of the same sort were known and utilized in explanation of various problems. Empedocles is said by Aristotle[106] to have accounted for the fact that the earth does not fall by referring to the observation that cups partially filled with water may be whirled round in a vertical circle without spilling the water —an illustration still used in text-books. It is perhaps doubtful whether Aristotle rightly understood his predecessor, who may rather have been accounting for the central position of the earth and the outer distribution of water, air, and fire in consequence of a rotary motion in a horizontal plane. As Empedocles adopted the early view that the segregation of the elements from the primordial chaos was the result of a vortex-motion, it is natural to suppose that it was by this process that he explained the origin of the heavenly bodies; for he is reported[107] as holding that the stars were fiery, and were forced out by the enveloping air in the primal separation. Anaxagoras[108] similarly held that the enveloping fiery aether, owing to the force of its revolution caught up rocks from the earth's surface, which took fire and turned into stars. Without wishing to defend these theories one may be sure that the Greeks had experimented with centrifugal forces and pondered on their possibilities.

Were one to include the practical arts and the details of technology there would be no limit to the evidences of experimentation; for it is obvious that progress in these fields

[105] 12, 852 a 38 ff. The principle is that of the lever. The *onagros*, an engine of war, applied it to missiles of great weight. See Diels, *Antike Technik*,[2] p. 98 f.

[106] *De Caelo*, II. 13, 295 a 16 ff.

[107] Aëtius, II. 13, 2 (Diels, *Doxographi Graeci*, 341).

[108] *Ibid.*, II. 13, 3 (*ibid.*, 341).

depended chiefly on the tentative steps taken in the effort
to attain more satisfactory results.   Prehistoric archaeology is teaching us daily to seek the origins of many arts
earlier in time, and consequently we have only the mute material witness of artifacts regarding the processes involved
in their making.   Even among the Greeks, where we have
a comparatively rich literary tradition, the data, absolutely
considered, are entirely inadequate to give a real insight
into the procedure of the artisans.   Our present concern,
however, is with science and not with the industrial arts,
except as they lent themselves to scientific purposes or led
to the formulation of generalizations and theories.   Something at least clearly akin to the scientific procedure must
be recognized in the *Canon* of the sculptor Polyclitus in
which he set forth the ideal proportions of the human body.
One can not doubt that observation and experiment laid
the foundation for his theory, which he expressed in mathematical terms and illustrated in typical statues.   To the
same category we must refer the study of the ideal proportions of Greek temples, though we are compelled to seek the
formulas evolved by the architects in the course of time in
the remains of their structures and in the late treatise of
Vitruvius.   Whether or not we accept Mr. Jay Hambridge's
"dynamic symmetry" of the Greek vase, a study of the
masterpieces of the Greek potters must convince us that
more than unsystematic observation and experiment was
required in the making of them.   How much well-consid-
ered theory resulted from the practice of such arts and was
transmitted from master to apprentice we can only guess,
because the tradition was doubtless oral.   The same is true
of the other arts.   One has only to call to mind the bridges
and aqueducts of the ancients in order to realize that experiment and theory must have gone hand in hand; and
much of this theory undoubtedly bore a distinctly scientific

aspect, though it originated in the solution of practical problems, as for example the observation of Frontinus that the rate of flow from a water-pipe depends upon its bore and the level of the supply in the reservoir.[109]

It requires little intelligence to perceive that fire-arms, not to speak of long-range cannon, call for considerable scientific knowledge in their manufacture. We should infer with certainty that somewhat the same order of knowledge was demanded for the making of ancient artillery, even if we had no literary documents to prove it. The fundamental laws of statics and ballistics must have been at least empirically observed, even if they could not be definitely formulated; and experiments must have been constantly made with a view to attaining a maximal efficiency. It happens that we still have certain writings on the subject which enable us to show that the *a priori* inferences are something more than guesses. One has only to read the introductory paragraphs of the treatise of Philo of Byzantium[110] on artillery to learn how close was the connection between experiment and theory in the production of the ancient engines of war. In Heron's similar treatise there are few measures given: in Philo and Vitruvius all parts of the catapults and ballistae are calculated in terms of the fundamental unit, the caliber of the bore. Since these specifications are asserted to be based on observation of the engines which in practical use proved to be most efficient, we have in effect an experimental science.

The fact seems to be that in the later centuries of ancient Greek history experiments in physics and other branches of science were being constantly made. If scientists in

[109] Frontinus, *De Aquis Urbis Romae*, I. 35.

[110] I have used *Philons Belopoiika*, ed. H. Diels und E. Schramm, *Abhandlungen der Preuss. Akademie der Wissenschaftem*, 1918. *Philos.-Hist. Klasse*, No. 16, Berlin, 1919.

modern times have been slow to recognize the truth in this
regard the fault lies partly in their absorption in their own
particular problems and partly in the deplorable loss of the
most important scientific literature and the break in the
oral tradition handed down in ancient times from master
to pupil. The causes which brought about this situation
are too well known to require further comment. The discerning reader will nevertheless readily understand the
procedure of the ancients, if he has learned the art of reading between the lines and asking the question, which must
constantly obtrude itself, how the author comes to make
the statement or present his thought in the way he does.
In Galen, for example, one may be so wearied by his
a priori reasoning that one fails to recognize the knowledge
of fact upon which he bases his deductions; yet patient
attention will reveal a wealth of detailed information
gained by observation, dissection, and experiment, which
may have to be gleaned from wearisome arguments. The
preference of the Greeks for logical deduction tends to make
the modern reader sceptical of their methods, especially
when the conclusions they drew differ from those which are
now accepted. If one wishes to correct the hasty judgment
which one is only too likely to form of the procedure of the
Greeks, there is perhaps no better text to read with care
than the introduction to the *Pneumatics* of Hero;[111] for the
experiments there suggested are so numerous and follow
so rapidly one after another, that one has the impression
of reading a laboratory manual. Diels[112] has suggested
that this statement is almost textually taken from Strato,
and this may in fact be true; for the central doctrine which
Hero set himself to establish is unquestionably that which,

[111] *Heronis Alexandrini Opera*, ed. Schmidt, I. 9–28.
[112] "Über das Physikalische System des Straton," *Sitzungsber. der Preuss. Akademie der Wissenschaften*, 1893, 101–127 (text, pp. 120–127).

as we know from other sources, Strato taught. But it would obviously be a mistake to conclude, as certain notable scholars have concluded, that the method of enforcing a point in an argument by repeated experiments was peculiar to Strato, who, because of his devotion to natural science, was known as the "Physikos." The passage in question is too long to transcribe; but a glance at the list of experiments included in it shows that practically all of them were known to the naturalists of the Fifth Century B.C., as has been pointed out in our foregoing account. The inference which one naturally draws is rather that we are here dealing, as in a modern laboratory manual, with stock illustrations, which were familiar to all students. The concern of Hero (or of Strato) with these experiments was solely that of establishing the doctrine that an absolute, continuous void does not exist in nature, but that a relative void, intercalated between molecules of matter, does exist and may be artificially created by various devices.

The tale of experiments enumerated in the foregoing account might be multiplied many times if it were desirable to catalogue them all. Those which have been mentioned are, however, sufficient to afford a just view of the procedure of the Greeks. How such a survey may affect the general reader who is interested in science must be left to others to say; to the writer, accustomed by years of study to the thought that the Greeks in all that they did merely exemplified the free working of human intelligence urged on by an insatiable curiosity, the showing appears to be just what one might have expected. If the scientist of today is most impressed by the obvious limitations of the Greeks and is inclined to belittle their achievements in comparison with those of the most recent generations, he may be properly reminded that the Greeks had to begin virtually *ab initio* and build the entire structure of science for themselves. It is not often given to the pioneer in any field of human endeavor to lay the foundations and to complete the edifice. As for science, one suspects and hopes that it will never be completely achieved.

As regards their experiments, one must acknowledge that for the most part they were simple and served to answer simple questions. Apparently, but by no means certainly, they were undertaken to test theories rather than to establish facts upon which to found theories. The distinction is often made, but one may be pardoned for doubting whether it is significant, if indeed it is valid. In contrast to modern science, the use of experimentation to break up a complex phenomenon into its component elements was rare, if not unknown, in ancient times. In like manner the quantitative relations, which can be determined only by experiment, were but rarely investigated and laws capable of mathematical statement were formulated in few fields. Consequently the advantage of carrying on extensive experiments under varying conditions and with different but related materials was virtually unknown to the Greeks. Since modern science owes its triumphs largely to these and similar methods one recognizes a great difference between it and the achievements of the Greeks. But, again, this is precisely what was to be expected. Methods and techniques are developed in response to the demands of investigation. The pioneers were scouting, as it were, surveying the field of experience in general; only as inquiry was pressed farther and farther, bringing new facts and unsuspected relations to light, could questions be framed that called for the application of further tests. What the Greeks did in the experimental sciences was to lay sure foundations in theory and practice. They made a beginning and pointed the way for those who should follow. One may say of them, in the words of Scripture, "And these all, having had witness borne to them through their faith, received not the promise, God having provided some better thing concerning us, that apart from us they should not be made perfect." Modern science is the consummation to a degree of the hopes and ideals that inspired those who first attempted the conquest of nature by human intelligence.

VI. EPILOGUE

In the foregoing chapters the effort has been made to illustrate by examples, drawn generally from the broad field of Greek thought, the conception, ideals, and methods of science. Controversial questions have been purposely avoided, and therefore many things which might have been mentioned were excluded. A history of science, in particular a history of Greek science, is a desideratum, but it may well be doubted whether the time has arrived for such an undertaking. In order to do full justice even to the lesser task a combination of talents is required which few men possess, and an extent of precise knowledge which could be found only in a large group of specialists and coordinated only by the closest cooperation. Something less than a complete survey would undoubtedly have a great and lasting value, and in certain limited provinces of knowledge excellent preliminary studies have been carried far; but one may perhaps question whether they do not discourage, rather than encourage, further inquiry. A great deal depends on how one conceives of the nature of science. If it be a body of ascertained facts fully certified and arranged under laws or generalizations culminating in universal law, the structure is indeed imposing, but it is not unlikely, as in the case of Aristotle, to paralyze the student who contemplates it and to render him impotent to press his search. A few daring intellects will, of course, venture farther, but the majority, possessing talent rather than genius, will probably confine their efforts to minor details. But if science be essentially a spirit or an attitude of mind which may express itself in different ways and apply itself to the understanding of any phase of nature and of life, there is abundant scope

195

for every talent and an incentive for its employment.   Certainly the latter point of view more nearly represents the Greek approach to problems.   Perhaps it will be said that this is the attitude of philosophy rather than of science; but it accounts not only for the development of science among the Greeks, but also for the easy passage, as for example in Plato, from the consideration of the fundamental problems of mathematics or epistemology to the practical tasks of statecraft.   Perhaps this explains the generally high regard for "philosophers" in Greece, and the frequency of their employment in affairs of public concern, whether as engineers, as "founders" of colonies, as legislators, or as ambassadors.

In this sense science appears to be peculiarly the creation of the Greek mind; and it is characterized by a singularly free play of the intellect, no matter to what subject it addressed itself.   The conception of political or civic liberty in an effective way may rightly be credited to this extraordinary people; but they did not achieve a single permanent state based on unalterable principles.   Rather they were by nature experimentalists; and the vast number of Greek city-states, with their different and changing constitutions, may well be likened to a laboratory or proving-ground for the testing of political theories.   Certainly Plato founded his theories upon the experience gained in this way, and the comprehensive studies of Aristotle in this field are responsible for the permanent value of his *Politics*.

In the physical sciences, the value of the work of the Greeks has been variously estimated according to the point of view of the historian, and this in turn has largely depended on his particular interest.   The judgment of the older historians was generally unfavorable.   In most cases one readily recognizes the reason for their disapprobation, which is founded on the supposition that Aristotle was the representative Greek thinker.   The natural and altogether wholesome revolt of the Renaissance against the dogmatism of the intrenched Aristotelianism called forth battle-cries that still reverberate in certain cloisters to which rumors of modern research scarcely penetrate. Some sayings of Whewell have perhaps done more than anything else to discourage scientists from studying ancient writers, as when he said of the Greeks,[1] "The defect was, that though they had in their possession Facts and Ideas, the Ideas were not distinct and appropriate to the Facts." This may apply to Aristotle's notions of potentiality and actuality; but to give it as a deliberate judgment on Greek science, even physical science, as a whole, is to condemn oneself as incompetent to exercise the functions of an historian.

To what extent the Greeks were original and how much they may have learned from other peoples are questions of fact, which are now, and probably will always remain, impossible to answer. Answers have, of course, been given, but they depend largely on prejudice. A distinguished historian of mathematics has said,[2] "For the mathematician the important consideration is that the foundations of mathematics and a great portion of its content are Greek. The Greeks laid down the first principles, invented the methods *ab initio*, and fixed the terminology. Mathematics in short is a Greek science, whatever new developments modern analysis has brought or may bring." "The Greeks were not merely the pioneers in the branches of knowledge which they invented and to which they gave names. What they began they carried to a height of perfection which has not since been surpassed; if there are exceptions, it is only where a few crowded centuries were not enough to provide the accumulation of experience required, whether for the purpose of correcting hypotheses, which at first could only

[1] *History of the Inductive Sciences*, I. 81.
[2] Heath, *A History of Greek Mathematics*, I. Preface, p. v.

be of the nature of guesswork, or of suggesting new methods
and machinery."[3]   If these statements claim too much for
the Greeks, scholars in general would concur in the conclusion of a French savant:[4]   "Never, so far as we know, in
all its centuries of existence, and even after coming into
contact with the science of the Greeks, did Oriental science
go beyond utilitarian interest and curiosity about details,
to rise to pure speculation and the determination of principles."   And yet we may not be too certain even of this
inference; for the decipherment and interpretation of Babylonian and Egyptian documents are bringing forward an
increasing body of evidence which calls for critical evalua-
tion.[5]   A layman in these matters may be pardoned for
believing that the criticism of Oriental documents has not
yet attained a sureness of method sufficient to enable the
general historical student to accept without question its
conclusions; but there seems to be no doubt that definite
data are being produced which the future historian must
seek to assess with an open mind.   Whatever the final
judgment of their intrinsic worth, the question will still
remain whether, or to what extent, the Greeks knew and
profited by the discoveries of their Oriental neighbors.   The
impression one gets from a study of Greek science is certainly that it is essentially independent of external influences,
although the eager Greek took his information about matters of fact wherever he found it.   He certainly did not
conceal his debt to the Orient; on the contrary, there is
every reason to suspect that in many cases he invented a
theory of borrowing from abroad, where there was none in

[3] *Ibid.*, I. 1.

[4] Léon Robin, *Greek Thought and the Origins of the Scientific Spirit*, 32.

[5] See the survey of R. C. Archibald, "Mathematics before the Greeks,"
Science, Jan. 31, 1930; *idem.*, "Babylonian Mathematics," *ibid.*, March 28,
1930, p. 342.

fact. Research has revealed a great store of common belief and practice existing among peoples who have had no demonstrable connection in historical times. The Greeks, with fewer data relating for the most part to peoples whom they knew more or less at first hand, naturally assumed borrowing in such cases, and produced theories to account for it. We have learned to be less certain of the ways in which different peoples arrive at the same result; even here it is well to be reminded of the logical principle of the plurality of causes.

These questions, however, do not properly concern us here, because this study does not purport to present a history of scientific thought. The present ambition of the writer will have been satisfied, if the reader obtains from it an impression of the way in which the Greeks, trying to introduce order into the apparent chaos of experience, used the native resources of the intellect to create a technique for theory and practice. The Greek seems to have felt, as did Wordsworth, that "the world is too much with us;" its very jostlings gave him a sense of being an alien until he could, as it were, keep it at arm's length long enough to glimpse its meaning. Its significance and relations fascinated him—if he could discover these, the brute facts interested him little. That many of his guesses went wide of the mark, means only that he was human; that he returned again and again to the attack, and never gave up the attempt to read the hidden meaning of the world by the light of his limited experience, proves that he possessed the spirit of the scientist and the philosopher. Once one realizes this irrepressible urge of the ancient Greek, his every enterprise acquires an interest for the thoughtful student, who values the idea more highly than the material in which it may chance to be embodied. Where the pioneers with the light heart of youth and inexperience thought to clear at a leap abysses which the ages have not sufficed to bridge, one must have grown old indeed if one fails to admire their adventurous spirit.   May it not be that in that spirit, informing everything they attempted, there is to be found the richest legacy which a highly endowed race has bequeathed to the modern world?

INDEX

Abdera, 108
Aeschylus, 54
Agatharchus, 183
age, new, 44
analogy, 67, 86, 139 ff.
analytical method, 100
ananké, 24
Anaxagoras, 183, 188
Anaximander, 29, 122, 123, 128
Anaximenes, 119, 121, 166
Arcesilaus, 111
Archimedes, 100, 171
architecture, 184
Archytas, 144 f.
Aristarchus of Samos, 99.
Aristotle, 1, 45, 68, 180, 185 f.
    his theory of induction, 95 f.
    his relation to medicine, history,
        and geography, 29
Aristoxenus, 52
astrolabe, 129
astrology, 41
astronomical instruments, 128 f.
astronomy, 128 ff.
    Babylonian, 129
Atomists, 38, 108 f.
auscultation, 179
Bacon, Roger, 78
ballistics, 190
capillarity, 181
Carneades, 111
Caspian Sea, 26
cause, 34 ff., 40
causes, plurality of, 110
centrifugal force, 187 f.
chance, 40, 43
chronology, 130 ff.
Cidenas, 129
classification, 118 ff.

clepsydra, 124, 167 f., 173 f.
Cnidus, medical school of, 136 f.
communicating tubes, 173
Cos, medical school of, 136 f.
cosmic ages, 44
cosmogonies and cosmologies, 7
creation, no, *e nihilo*, 13, 38
creation as procreation, 13
cupping-glass, 147
Darius Hystaspes, 26
definition, genetic and teleological, 10 ff.,
    14, 24 f.
Demetrius, 112 f.
Democedes, 26 f.
Democritus, 108, 183.
Dicaearchus, 29 f., 98, 106.
dichotomy, Platonic, 138
diké, 24
Diogenes of Apollonia, 142
dip of the earth, 122
disk-earth, 122 f.
dissection, 140 f., 178
doxography, 27 ff.
earth, disk and globe, 98, 122 f.
Eleatics, 37 f.
element, 36.
elements, four, 119 f.
elimination, method of, 138.
embryology, 180
Empedocles, 38, 119 f., 188
Empirics, their doctrine of induction,
    113 f.
Epicharmus, 51
Epicurean theory of induction, 112 ff.
Epicurus, 43, 47, 110.
equator of early maps, 125.
Erasistratus, 109, 176, 180.
Eratosthenes, 29, 53 f., 98, 126
eschatology, 44

201